WHOLE

BRAIN

LIVING

ALSO BY

JILL BOLTE TAYLOR, Ph.D.

My Stroke of Insight:
A Brain Scientist's Personal Journey

WHOLE
BRAIN
LIVING

the ANATOMY *of* CHOICE
and the FOUR CHARACTERS
THAT DRIVE OUR LIFE

JILL BOLTE TAYLOR, Ph.D.

HAY HOUSE, INC.
Carlsbad, California • New York City
London • Sydney • New Delhi

Published in the United States by: Hay House, Inc.: www.hayhouse.com®
Published in Australia by: Hay House Australia Pty. Ltd.: www.hayhouse.com.au
Published in the United Kingdom by: Hay House UK, Ltd.: www.hayhouse.co.uk
Published in India by: Hay House Publishers India: www.hayhouse.co.in

Cover design: Jason Gabbert
Interior design: Julie Davison
Interior illustrations: Merridee LaMantia
Indexer: J S Editorial, LLC

Cataloging-in-Publication Data is on file at the Library of Congress

Hardcover ISBN: 978-1-4019-6198-5
E-book ISBN: 978-1-4019-6199-2
E-audio ISBN: 978-1-4019-6200-5

10 9 8 7 6 5 4 3 2 1
1st edition, May 2021

Printed and bound in Great Britain by
TJ Books Limited, Padstow, Cornwall

I am forever grateful for the love
of GG and Hal, Florence and Bill,
and Poppy and Dandy.

CONTENTS

PREFACE

Peace Is Just a Thought Away

In 2008 I received an invitation to give a TED talk. At that point in time, there were only six TED talks online and I had no idea what TED was. (It turned out to stand for Technology, Entertainment, and Design.) The TED talk "My Stroke of Insight" that I gave in Monterey, California, was the first to ever go viral on the Internet. As a result TED and I became globally famous simultaneously.

In that talk I shared with the audience my story of surviving a massive cerebral hemorrhage in which the left hemisphere of my brain shut down and the right hemisphere became dominant. I described how I, through the eyes of a neuroscientist, watched with fascination as my circuits and faculties went off-line. I took the audience on a journey into the deterioration of my own left brain, whereby I shifted into a state of peaceful euphoria and oneness with the universe, unlike anything I had ever known before.

Within three months of delivering that talk, I was chosen as one of *Time* magazine's 100 Most Influential People in the World for 2008. I became the premiere guest on Oprah's Soul Series webcast, and my memoir, *My Stroke of Insight: A Brain Scientist's Personal Journey*, was published by Penguin Books. It spent 63 weeks on the

New York Times bestseller list. Today, over 12 years later, it remains the number one book in the Amazon marketplace on the subject of stroke, and in the top ten in other subjects including Anatomy Science, Medical Professional Biographies, and Nervous System Diseases.

That 18-minute presentation instantly changed my world, and many other lives were permanently shifted too. I still have TEDsters come up to me to share what row they were sitting in on that fateful afternoon, and over 25 million views later, that talk remains one of the most popular TED talks of all time. I have received hundreds of thousands of emails over the years from people asking how they, too, can access that peaceful euphoria I described.

Unquestionably, in many ways that TED talk was an amazing success.

But in my heart the talk failed to accomplish one thing I had hoped it would do. I wanted us, as human beings, to recognize that we are connected as part of a whole, and I wished for us to treat one another with a higher degree of respect and kindness. Instead, our civility toward one another has clearly decayed over the past decade and more.

Perhaps this is not so surprising, given that we live in a world where our politics, relationships, and life in general are spiraling into an uncomfortable state of chaos. Life is tough with its highs and lows, and none of us came into this world with a manual for how to get it all right. But what I have learned is that we do have the power to pause, sidestep our habituated patterns, and make better choices. We have the power to choose moment by moment who and how we want to be in the world.

That ability, like every ability we have, is dependent on the cells in our brain that perform that function. Our brain is a magnificent tool that is the home of our thoughts, emotions, experiences, and behaviors. When we understand at a cellular level what is going on in the relationship between our thoughts and our emotions, we no longer have to be bound by our emotional reactivity. Instead, we can become emboldened to live our best lives and be our best selves. We have much more power over what is going on inside of our heads than we have ever been taught.

Much of the material in this book stems from my own experience with watching my brain break down due to trauma, and then insights I gained as the cells in my brain recovered. In the big picture, this book is about our shared journey into the challenges of our lives and what our choices are in how we can live our best life while taking our brain anatomy into account. In these pages you will encounter a new paradigm for understanding how the different parts of our brain work together to manifest our perception of reality, along with a set of tangible tools you can use to not only master your brain's emotional reactivity, but ultimately live a whole-brain life.

You are the life-force power of the universe and your human brain is amazing far beyond your wildest imagination. This book will clarify for you exactly what that means and what your options are as you learn to own your power and apply these tools to create the life you want to live. This book is your road map to peace, which really is just a thought away.

Part I

A BRIEF LOOK INSIDE YOUR BRAIN

CHAPTER 1

My Story and Our Brain

I grew up to study the brain because I had a brother 18 months older than I who would eventually be diagnosed with the brain disorder schizophrenia. As young siblings, my brother and I were virtually inseparable, but at an early age, I realized that he and I were very different from one another in how we experienced reality. On a routine basis, we would have the exact same experience but walk away with very different interpretations of what had just happened. He might think our mother was angry at us based on the tone of her voice, for example, while I was quite sure she was petrified that we were going to be hurt. Because of this I became fascinated with trying to understand what was "normal" because it was clear to me that one of us was atypical. As far as I could tell, he was oblivious to our different perceptions and interpretations.

For my own survival and sanity, I started paying very close attention to what I could learn from others based on their body and facial languages. I became fascinated with anatomy, and at Indiana University I pursued undergraduate degrees in physiological psychology and human biology. After spending a couple of years as a lab tech in a neuroanatomy lab, I skipped the master's degree program and went straight for a Ph.D. in life sciences at Indiana State University.

Although my research focus was in neuroanatomy through the Indiana University School of Medicine, I found my true joy in the gross anatomy lab, where we dissected human cadavers. For me there is truly nothing more magnificent than the human body, so "gross" lab was a spectacular treat. It was during this doctoral program that my brother, at the age of 31, was officially diagnosed with chronic schizophrenia. As you might imagine, a part of me felt relieved to learn that he was the one diagnosed as "not normal," as that meant that I was most likely the neurotypical one.

After I received my doctorate in Indiana, I scooted off to Boston, where I initially spent two years in the Harvard Department of Neuroscience. From there, I spent four years in the Harvard Department of Psychiatry, working with the amazing "Queen of Schizophrenia," Dr. Francine Benes. My research and professional life truly began to blossom. I adored being a lab rat and felt an awe-inspired camaraderie with the beautiful cells I examined through the microscope.

I was fascinated with how our brains create our perception of reality. I studied the postmortem brain cells and circuitry of people who were diagnosed as normal-control—meaning they would be used as the control group in the experiments I was designing—and then compared that tissue with the brains of individuals diagnosed with schizophrenia, schizoaffective, or bipolar disorder. My weekdays were spent performing jaw-dropping innovative research, which ultimately resulted in journal articles with titles like "Differential Distribution of Tyrosine Hydroxylase Fibers on Small and Large Neurons in Layer II of Anterior Cingulate Cortex of Schizophrenic Brain" and "Colocalization of Glutamate Decarboxylase, Tyrosine Hydroxylase and Serotonin Immunoreactivity in Rat Medial Prefrontal Cortex." This last one became a classic, as it was the first article ever to be published by the first online-only scientific journal, *Neuroscience-Net*.

On the weekends, with guitar in tow, I took a different tack. I traveled as the "Singin' Scientist" for the Harvard Brain Bank, educating families with mental illness about the shortage of brain tissue for research and the value of brain donation. At the age of 36, I found myself the youngest person to ever be elected to the national board of directors of NAMI, the National Alliance on Mental Illness. This wonderful organization has a membership of over 100,000 families whose loved ones have been diagnosed with severe mental illness. NAMI is a really important national, state, and local resource for families in need (NAMI.org). Between the research and advocating for the mentally ill at the national level, my life had terrific purpose. I was helping people like my brother while at the same time keeping my finger on the pulse of research and public policy.

I was in the prime of my life, strong and athletic, and climbing the Harvard ladder. I was fulfilling my dreams as a successful neuroscientist in the world of schizophrenia and finding meaning as a national-level advocate. Then, on the morning of December 10, 1996, at the age of 37, I woke up to a pounding pain behind my left eye.

My Stroke and Insight

As it turned out, I was born with a congenital neurological brain disorder that I didn't know was there until it became a problem. An arteriovenous malformation (AVM) exploded in the left hemisphere of my brain, and over the course of four hours I watched my brain functions shut down one by one. On the afternoon of the stroke I could not walk, talk, read, write, or recall any of my life. In effect, I had become an infant in a woman's body.

As you might imagine, it was fascinating for me to watch my brain systematically break down, through the eyes of a neuroscientist. The damage to the left hemisphere of my brain was so traumatic that I predictably lost the ability to speak and understand language. In addition, the chattering "monkey mind" of my left brain went silent. With that internal dialogue circuitry shut off, I sat in the center of a completely silent brain for five full weeks. I even lost that little voice of my left-brain ego-self that could say, "I am an individual, separate from the whole. I am Dr. Jill Bolte Taylor." In the absence of my chatty and linear-thinking left brain, I stepped into the awe-inspiring experiential sensations of the present moment, and it was beautiful there.

Compounding my deficiencies of language and individuality, the injury to my left parietal lobe, which processes sensory information from the outside world, made it impossible for me to identify the boundaries of where my physical body began and where it ended. As a result my perception of myself became altered. Instead of a physical being, I experienced myself to be an energy ball that was as big as the universe. Shifted into this consciousness of my right brain, I perceived the essence of myself as enormous and expansive, and my spirit soared free, like a great whale gliding through a sea of silent euphoria.

Emotionally I went from feeling the normal emotions I had experienced in my pre-stroke life to feeling nothing except a sense of peaceful bliss. I know this sounds like an amazing blessing, and it certainly was, but being able to feel a range of emotions makes life much more diverse and interesting. Physically, within those same four hours on the morning of the stroke, I went from being able to swim a mile in 30 minutes to lying sprawled on a hospital gurney with my conscious mind trapped inside a motionless body that felt like a ton of lead.

It was eight years before my body completely recovered and I could slalom water-ski again. During that time I regained the emotional circuits of resentment, guilt, and embarrassment, as well as all of the other more subtle feelings and emotions that make life alluring. Our emotions, even the negative ones, truly enrich our perception of experience and make life nuanced and more remarkable. I wrote about this stroke, recovery, and lessons learned about neuroplasticity and the ability of the brain to recover in my memoir, *My Stroke of Insight: A Brain Scientist's Personal Journey.*

Since then I have begun to explore even more deeply the most valuable insight I gained from this sojourn into the depths of my brain: the realization that we have the power to turn our emotional circuitry on and off by choice. In fact, the same principle underlying our bodies' neurological reflexes, like the kick our knee will give when our patella tendon is tapped with a reflex hammer, remains in play when an emotional circuit is triggered and we reflexively respond with fear, anger, or hostility.

Once the circuit is stimulated and we have triggered an emotional response, it takes less than 90 seconds for the chemistry of that emotion to flood through us and then flush completely out of our bloodstream. Of course, we can either consciously or unconsciously choose to rethink the thought that triggered the emotional circuit to run and stay hurt, angry, sad, or whatever for longer than 90 seconds. But in that case what we are doing at a neurological level is restimulating the emotional circuit so it will run over and over again. If there is no repeated trigger, the emotional circuit will run its course and stop after the 90 seconds that it takes for the chemistry to neutralize. I call this the 90 Second Rule and will share examples in the chapters ahead.

The "We" inside of Me

The TED conference where I presented was dedicated to "The Big Questions," and for the opening session we speakers were directed to address the theme "Who Are We?" I chose to approach this by talking about the "We" inside each of our brains, the "We" of our left and right brain hemispheres. The roster of speakers included some world-famous scientists, including the Canadian anthropologist Wade Davis and Louise Leakey, a National Geographic paleontologist. Then there was me, a Harvard-trained girl from Indiana who had survived, and recovered from, a massive stroke. Needless to say, I was the least known speaker in the lineup.

We have the power to choose, moment by moment, who and how we want to be in the world.

On the day before the opening of the conference, I was onstage giving a practice run for the TED staff and crew. They were performing sound and lighting checks and working through logistics, and because I had brought along a preserved human brain, there were some special considerations. After delivering the first six minutes of my presentation during that practice run, I paused and was ready to stop, but Chris Anderson, the curator of TED, encouraged me to proceed. His mother had experienced a stroke, so he was particularly interested in my topic.

For the next segment, I took the audience on a reenacted journey into the moment-by-moment breakdown of my mind as it had occurred on the morning of the stroke. I shared what it felt like to waffle between the consciousnesses of my left and right hemispheres. It was a dramatic performance in which my left hemisphere was desperate to orchestrate my rescue, countered by the blissful euphoria of my right hemisphere.

I described how I struggled to stay connected to my functional left brain and managed to make a phone call for help, even though I had no recognizable language. When I found myself curled up in a fetal ball in an ambulance, I felt my spirit surrender, and in that release I was certain I was in my moment of transition. At that point in the presentation, to my surprise, an uncanny silence fell over the TED practice room, and I realized that the staff and crew had stopped what they were doing to listen.

Quote: "When I awoke later that afternoon, I was shocked to discover that I was still alive. When I felt my spirit surrender, I had said goodbye to my life. Then I realized, but I'm still alive, and I have found Nirvana, and if I have found Nirvana and I'm still alive, then everyone who is alive can find Nirvana. I pictured a world filled with beautiful, peaceful, compassionate, loving people who knew that they could purposely choose to step to the right of their left hemispheres and find this peace. And then I realized what a tremendous gift this experience could be, what a stroke of insight this could be for how we live our lives. And it motivated me to recover."

The room was not silent anymore. When I finished, I heard sniffles and even weeping. Chris immediately rearranged the schedule, shifting my talk to the last spot of the afternoon. I might be an unknown girl from Indiana, but he knew that this presentation was something special and that his attendees would probably be profoundly moved. It turned out he was right.

Thanks to the response of the staff and crew, I slept well that night and woke up fresh before taking the TED center stage. I ended my talk in this way, answering the "big question" with this reminder:

Who Are We?

We are the life-force power of the universe with manual dexterity and two cognitive minds. We have the power to choose, moment by moment, who and how we want to be in the world.

Right here, right now, I can step into the consciousness of my right hemisphere, where we are, I am, the life-force power of the universe. I am the life-force power of the 50 trillion beautiful molecular geniuses that make up my form, at One with all that is.

Or, I can choose to step into the consciousness of my left hemisphere, where I become a single individual, a solid, separate from the flow, and separate from you. I am Dr. Jill Bolte Taylor: intellectual, neuroanatomist.

These are the "We" inside of me.

Which would you choose? Which do you choose . . . and when?

I believe that the more time we spend choosing to run the deep inner peace circuitry of our right hemisphere, the more peace we will project into the world and the more peaceful our planet will be.

And I thought that was an idea worth spreading.

WHAT THIS MEANS FOR YOU

As I have already mentioned, the public's response to that TED talk continues to be profound. It is clear that we, as a collective, are searching for a specific set of directions about how we can choose the peaceful mindset of our right hemisphere to counterbalance the chaos in our world. Many of us are in search of a paradigm shift for how we can embrace our deep inner peace, regardless of our situation.

The most frequently asked question I receive is "How do I quiet the brain chatter in my left brain?" Clearly, many people want to stop their habits of self-judgment and criticism. It is also common for me to hear "I have been practicing meditation for years and have only experienced that feeling of euphoria that you describe a couple of times. What can I do differently to get there? Do you meditate, and if so, what form? Can you still find that sense of euphoria, and if so, what can I do to get there?" Then there is this one: "What drugs might I take to help me feel that euphoria you got from the stroke? Psychedelics? If so, which ones?" (As important as that question is, especially in light of the recent research into the use of ecstasy for post-traumatic stress disorder [PTSD], that field is beyond my knowledge base.)

By choosing to meditate, pray, or practice mindfulness exercises, we can certainly quiet the chatter and free ourselves from the prison of our own minds. But please be clear: This book is not about any of those subjects. It is about "the power of the We inside of me." I believe that the better we understand the various groups of cells inside of our brain, how they are organized, and what it feels like to run each of those different cellular circuits, the more power we have to purposefully choose which of those neural networks we want to run. By doing so we ultimately gain the power to choose who and how we want to be in the world each and every moment, regardless of what external circumstances we find ourselves in.

Peace really is just a thought away. It is always right there, and always available for you to embody.

I will draw on two different disciplines to explain this idea in the chapters ahead. The science of neuroanatomy involves the study of the structure of the brain. The science of psychology involves the study of the mind and our mental processes. What makes this book so unique and exciting is that the psychology I present here is specifically correlated to the underlying brain anatomy and what we know about the function of those specific groups of cells. When you open yourself to this material, you will gain amazing insights into both the conscious and unconscious realms of your left and right hemispheres. By doing so you will become much more aware of your power to choose who and how you want to be, because you will know what your options are at both a psychological and biological level.

This journey we will take into your brain is reminiscent of Joseph Campbell's classic monomyth that describes the steps a hero must take to fulfill his Hero's Journey. In the language of the brain, the hero must step out of his own ego-based left-brain consciousness into the realm of his right brain's unconsciousness. At this point the hero feels connected to all that is, and is enveloped by a sense of deep inner peace. As you gain mastery of the Four Characters in your brain, whom you are about to meet, you will embark upon your own Hero's Journey into the circuitry of your unconscious brain and realize that peace really is just a thought away. It is always right there and always available for you to embody.

When the cells in my left hemisphere experienced trauma and shut down, I did not just lose cells and skill sets. I lost parts of my personality, including the highly motivated part that was smart, disciplined, punctual, detail based, methodical, and well organized and that knew the details of my life. That part of me was a character that disappeared with the stroke and was no longer available, at least until those cells recovered and that circuitry came back online. I also lost the part of my personality that had known all of the challenges, emotions, and pain from my past. Without that character available, all I could experience was the peaceful euphoria of the present moment.

It took eight years for me to rebuild all of those wounded circuits, and then resurrect and recover those two left-brain characters that had gone off-line. I learned the hard way that we each have four distinctive groups of cells, divided between our two brain hemispheres, that generate four consistent and predictable personalities. Neuroanatomically these four groups of cells make up the left and right *thinking* centers of our higher cerebral cortex, as well as our left and right *emotional* centers of our lower limbic system. Collectively, I call these personalities the Four Characters. Getting to know them inside of your brain is a ticket to freedom.

I realize that the material in this book may require a theoretical shift in how you think about your brain anatomy. For at least 50 years, we have been trained as a society to believe that our left hemisphere is our "rational thinking" brain, while our right hemisphere is our "emotional" brain. Actually, from a neuroanatomical perspective, although it is true that our left thinking tissue is the home of our conscious, rational mind (which I will refer to as Character 1), both our left and right brain hemispheres share the cells of our emotional limbic system equally (Characters 2 and 3). Character 4 occupies the higher cortical thinking tissue of our right brain.

FOUR CHARACTERS

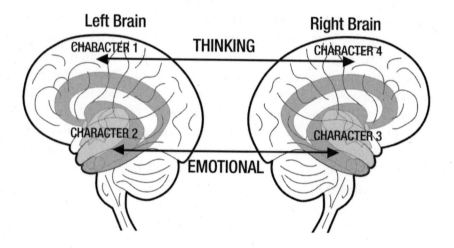

How We Think and Feel

At any moment in time, there are pretty much only three things going on in our brain. We think thoughts, we feel emotions, and we run physiological responses to what we are thinking and feeling. Each of these activities is completely dependent on the health and well-being of the cells that are performing those functions.

We experience emotions via the cells of our limbic system, and these cells are evenly divided between our two hemispheres. The major structures of the limbic system are mirrored in each hemisphere such that we have two amygdalae, two hippocampi, and two anterior cingulate gyri, among others. This means we have two separate modules for emotional processing (Characters 2 and 3). When information streams in through our sensory systems, it first stops off at our amygdalae, which are there to ask the question, "Am I safe?" We feel safe in the world when enough of the sensory stimulation coming in feels familiar.

TWO EMOTIONAL BRAINS

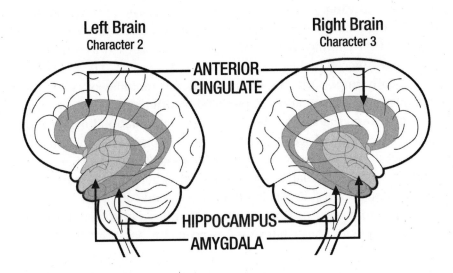

Left Brain
Character 2

Right Brain
Character 3

ANTERIOR
CINGULATE

HIPPOCAMPUS
AMYGDALA

When something does not feel familiar, however, our amygdalae tend to label that unfamiliar thing as dangerous, and they respond by triggering our fight-flight-or-play-dead fear response. If it has been your natural tendency to fight, you probably rage, get big and loud, go on the attack or try to shoo the thing away. If it is your style to run like the wind or play dead, then that response may be your best choice.

When our amygdalae are triggered and we feel fear, we are not able to run the learning and memory circuitry of our hippocampi.

Until we push the pause button and take a moment to calm down and feel safe again, we will not be able to think clearly. This is why anyone who is freaking out with test anxiety tends to perform poorly, regardless of how well prepared they are. When our limbic anxiety circuit is triggered, we are neuroanatomically cut off from accessing our higher cortical thinking centers, which is where our learned knowledge is stored.

Understanding the anatomy of the brain is always insightful when it comes to our experience and behavior. If we live with the basic belief that there is only one group of cells inside of our brain that processes our emotions, our experience of mixed emotions can be very confusing. At a neuroanatomical level, when we experience conflicting feelings it is because we have two emotional groups of cells that are completely separate from one another in that they do not share any cell bodies.

Equally important, these two emotional modules of cells process incoming information in predictably different ways. Providing we understand that our left brain processes information linearly and in sequence, we will see in detail how our left-brain emotional module is designed to bring in information about the present moment and then compare that to any emotional experience we have had in the past. As a result, our left-brain emotional Character 2 is programmed to protect us from anything that has a history of hurting us. Consequently, our Character 2 is primed to say "No" and push things away.

Our right-brain emotional Character 3 is exactly the opposite in that it processes present-moment experiences in the present moment. Therefore our emotional Character 3 always exists in the here and now and has no recollection of the past. Instead of pushing things away, our Character 3 moves enthusiastically toward any experience that remotely smells like an enticing and juicy adrenaline rush.

In the mammalian nervous system, a new species is often created by adding new brain cells on top of a well-integrated preexisting cellular matrix. When this happens, the new tissue is designed to refine and evolve the abilities of the tissue below. In the case of the human brain, although we share the cells of our deeper emotional limbic tissue with other mammals, such as dogs and monkeys, what distinguishes our human brain as unique are the newly added-on higher cortical cells of our two thinking brains.

When information from the external world streams in through our sensory systems, it is processed first by the cells of our limbic emotional cells before it is refined by our higher thinking centers. So, from a purely biological perspective, we humans are feeling creatures who think, rather than thinking creatures who feel. Neuroanatomically you and I are programmed to feel our emotions, and any attempt we may make to bypass or ignore what we are feeling may have the power to derail our mental health at this most fundamental level.

From an evolutionary standpoint, our human brain exists as a truly amazing neurological accomplishment, but it is critical to remember that we are far from being a finished product. Instead humanity exists in an ongoing state of evolution: First, we are actively integrating the newly added-on tissue of our left thinking brain (Character 1) with the tissue of the underlying left emotional brain (Character 2). Second, we are integrating the newly added-on tissue of our right thinking brain (Character 4) with the underlying tissue of our right emotional brain (Character 3). Third, we are connecting the left emotional brain tissue (Character 2) with the right emotional brain tissue (Character 3). And finally, we are integrating the left thinking brain tissue (Character 1) with the right thinking brain tissue (Character 4). When we accomplish this we will evolve into whole-brain living.

WHOLE BRAIN COMMUNICATION

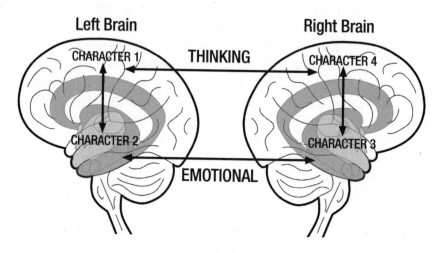

Left Brain

Right Brain

CHARACTER 1

THINKING

CHARACTER 4

CHARACTER 2

CHARACTER 3

EMOTIONAL

Although our human brain is an evolving masterpiece in process, you don't have to look far to see how the differences between what our left and right hemispheres value (which we will explore in detail in Chapter 3) are playing out in our lives and in society. Besides the most obvious social unrest imparted by our bipartisan political hostilities, statistically speaking, one in five adults in the U.S. will be diagnosed with a severe mental illness at some point in their life. Choosing to advance our own evolution as a species will help us find peace individually, communally, and ultimately globally.

As we move through this material, I implore you to open your heart and mind and be completely honest with yourself about your own individual strengths and weaknesses. As long as we live in a society that rewards us for what we do, rather than for who we are, we will feel undervalued and unfulfilled. For many of us, our goal has been to "get rid of" or "fix" the most unruly, unattractive, or vulnerable parts of ourselves. But when we choose to embrace, listen to, and nurture all of our characters, we will mature, grow, and evolve into that person our dog already thinks we are.

Just to clarify, in this book we are talking about four predictable and easy to identify characters that we all have, based on the anatomy of our brain. Every ability we have is completely dependent on the underlying brain cells that manufacture those abilities, and these four different groups of cells manufacture four different skill sets, ultimately resulting in the expression of each of our Four Characters. When many authors and teachers refer to the *authentic self,* you may wonder which of these Four Characters they might be referring to. In fact, from the way they describe the "authentic self," it is clear that they are referring to Character 4. But please understand that none of the Four Characters is more authentic than the others. Each of these characters represents an authentic part of who we are at a cellular level and should be treated with dignity, respect, and honor.

A Note on Brain Disorders

It is important to note that the material presented here is not related to either schizophrenia or multiple personality disorder, which are both serious neuropsychiatric disorders. The word *schizophrenia* by definition means "split brain," but the division that it describes is a break between the brain of a person and the accepted norms of its surrounding society.

The necessary criteria for someone to be diagnosed with the brain disorder schizophrenia is the experience of sensory hallucinations supported by a delusional thinking system. If your brain is inputting abnormal sensory perceptions of the world because you are seeing, smelling, or hearing things that others are not experiencing, it is impossible for your brain to use those building blocks to construct a normal perception of the world. Predictably, your brain would create a delusional thinking system matching the altered input. Not only do the brains of people with schizophrenia process incoming data in error from normal perception, but there are alterations in the internal wiring of that information. As a result, the brain of a person with schizophrenia is split from normal information processing at a cellular level, and the resultant delusional thinking system is a by-product of that brain's abnormal neurological miswiring.

Multiple personality disorder (MPD) is a completely different brain disorder from schizophrenia. There is a lot that is not known about this disorder, including why or how a brain is capable of manufacturing multiple personalities. Sometimes these personalities don't even know about one another, or they may exist in conflict. MPD is a pathological condition that may be manifested as a coping tool in response to childhood trauma. In the case of MPD, the split in consciousness occurs within the brain, whereas in schizophrenia the split occurs between the consciousness of the brain and its perception of external reality.

After the stroke, once my whole brain came back online and all four of my characters became fully functional again, I learned that I have the ability to not only recognize which circuitry or character I am running but to choose whether I want to continue running that circuit or switch to a different one. This unusual journey has helped me understand that not only I but all of us have incredible power concerning who and how we want to be. My passion is for you to master your own Four Characters so you, too, can completely own your power and live your best life.

In the following chapters we will explore the anatomy and psychology of the brain's two hemispheres and Four Characters in more detail. (Don't worry—I will make this as interesting and simple as I can.) From there we will explore the unique skill sets that each of the Four Characters specialize in and help you identify which character you are inhabiting at any moment based on what they *feel* like inside of your body.

Then, as we move farther into the book, not only will you meet and get to know the Four Characters of your left and right *thinking* Characters 1 and 4, and your left and right *emotional* Characters 2 and 3, but you will gain insight into how these Four Characters can interact and work together on your behalf.

When we know, understand, and nurture our own Four Characters, their relationships with one another, and their collective power within us, we promote our own cognitive, emotional, physical, and spiritual wellness. This is whole-brain living. I truly believe that this is the evolutionary goal of humanity, and we are getting there one brain at a time.

CHAPTER 2

Brain Anatomy and Personality

My father, Hal Taylor, was a preacher man. He was an ordained and practicing Episcopalian minister during my childhood, and when I was a teenager he became a therapist after receiving his Ph.D. in counseling psychology. Hal was fascinated with people of all walks of life and earned his living helping corporations and non-profit organizations develop team-building skills for better board management and performance. He did this using personality profiles and temperament typing.

Hal was obsessed with helping people help themselves, be they presidents of organizations, the severely mentally ill, or those incarcerated in our jails. Hal had a heart of gold, and his single goal in life, in my opinion, was to help people better understand their strengths so they could live more fully. Temperament typing was a great tool for this. His primary tool was the Myers-Briggs Type Indicator, which was very popular back in the 1970s, '80s, and '90s and is still being used by over a million people a year.

The first time Hal gave me the Myers-Briggs, I was 18 years old and starting college. Like many others I rebelled against the forced-choice nature of the exam because my answers were completely dependent on the circumstances in which I pictured myself. I originally tested as an INTJ—an Introverted, Intuitive, Thinking, and Judging personality. This profile, labeled by psychologist and temperament-typing expert David Keirsey as the Scientist, clearly depicted a character inside of me, but I was that character only part of the time. When hanging out with my friends, I was an ESFP Performer type—Extroverted, Sensing, Feeling, and Perceiving. So much so that I was voted class clown in high school.

The Myers-Briggs did not accommodate for different life scenarios, and because it pigeonholed me into a single character, I questioned the accuracy of the evaluation. This sparked in me a lifelong curiosity and drive to find a psychological typing system that was more anatomically accurate. Following in my father's footsteps, I became fascinated with psychology and the brain, as well as with the relationships between our mind, brain, body, and behavior. I loved anything that was human-based biology.

The Split-Brain Experiments

To my good fortune, I was an undergraduate during the late '70s, when the field of neuroscience went mainstream and the famous split-brain surgeries captivated the public's attention. To say the least, I was riveted by the work of Dr. Roger Sperry, who surgically separated the two cerebral hemispheres of several of his epileptic patients.

When Dr. Sperry surgically cut the corpus callosum—the band of some 300 million axonal fibers connecting the two cerebral hemispheres—in a procedure called a commissurotomy, he was successful in preventing dangerous seizure activity in one hemisphere from spreading into the other. There was another benefit too: the psychological experiments that were performed on this patient population by Dr. Michael Gazzaniga bore great insights into how our two brain hemispheres function differently when they are separated.

As a budding neuroscientist, I was particularly fascinated by the Dr. Jekyll and Mr. Hyde stories from these experiments, which depicted the psychological and underlying anatomical abilities of the two cerebral hemispheres as dramatically different. It was clear that when the two hemispheres were separated, the split-brain patients behaved as though they were two unique characters that often acted in direct opposition to one another.

In some of these patients, the character "occupying" the right brain hemisphere would directly contradict the intention and behavior of the character "occupying" the left brain hemisphere. For instance, when one gentleman attempted to slap his wife with his left hand (right brain), he simultaneously protected her with his right hand (left brain). On another occasion this same fellow was clearly conflicted when he yanked his pants down with one hand while simultaneously redressing himself with the other.

A different patient, who happened to be a child, was completely verbal in both of his hemispheres. When asked about his life goals, his right-brain character reported that he wanted to grow up to be a race car driver, while his left-brain character was interested in becoming a draftsman. Still another commissurotomy patient reported that she went to battle with herself every morning when choosing her clothing. She described her right and left hands as two repelling magnets, each of which had different styles in mind about what she should wear that day. The same thing happened when she was shopping for food at the grocery store, as her two hemispheric characters were interested in completely different cuisines. It took well over a year following her commissurotomy surgery before she was able to master a single intention and purposefully inhibit the internal battle going on between her two differently opinionated hemispheric characters.

As you hear these stories, it is important to note that the only anatomical difference between these commissurotomy patients and you and me is that our two cerebral hemispheres communicate with one another via the connections of our corpus callosum. Scientists understand that neuroanatomically the majority of these commissural fibers are inhibitory in nature and that they run from one set of cells in one hemisphere to the comparable set of cells in the opposite hemisphere. At any moment in time, both hemispheres have cells that are active, but opposing hemisphere cell groups dance between dominance and inhibition.

In this way, one hemisphere has the power to inhibit the function of the comparable cells in the opposite hemisphere, dominating the function of that particular group of cells. For example, when we are focused on the words and meaning of what someone is saying (left brain), we tend to not be so focused on the inflection of their voice or the emotional content (right brain) of what they are communicating. Vice versa—have you ever been so stunned that someone was yelling at you that you completely missed the point they were trying to make?

Back in the '70s and '80s, society went a bit overboard in its response to the split-brain studies, and all sorts of "right brain" and "left brain"–based community programs popped up. Many schools even got into the game and established curricula that would help stimulate one or both of the hemispheres. The stereotypes of left-brain and right-brain people entered the mainstream, with the left-brainers appearing to be more organized, punctual, and good with details while the right-brainers thrived in creativity, innovation, and athletics.

Unfortunately, in response to the left-brain/right-brain craze, the strategy that many parents took to help their children get ahead was to expose them to programming that fit their natural dominance. This makes sense, of course, since they wanted their children to be rewarded for what they did well. But if their goal was to create more rounded, whole-brained children, a better plan might have been for them to encourage their kids to partake in activities at which they did *not* excel. For example, they could have encouraged the left brain–dominant science and math types to participate in outdoor events in which they could explore and collect data in the woods. And they could have enticed the athletic and artistic types to creatively design some really cool science-fair projects that would measure some type of performance.

Because parents did it the way they did, however, over the last 40 years we have skewed our abilities toward the two extremes. There have been some writings and teachings specifically designed to help develop our nondominant side, including the book *Drawing on the Right Side of the Brain*, which is a classic and still widely used today. Also, you don't have to look far to recognize how marketers have mastered their advertising strategies to target our right- or left-brain preferences. Even our computer systems fit the bill: Apple products are viewed as right-brain creative, while anything Windows based screams left-brain analytical. Remember that Blackberry? It used to make my right brain moan.

How the Two Hemispheres Function

In addition to these pop-science efforts that have been designed to capitalize on the stereotypical differences between our two hemispheres, a tome of evidence-based science now offers us a pretty clear understanding of both the anatomical and functional differences between the two halves of our brain. For anyone interested in both the big picture and the details about what we have learned concerning these differences over the last half century, *The Master and His Emissary*, a book by the British psychiatrist Dr. Iain McGilchrist, is a fascinating and up-to-date read.

In addition, if you are interested in how a Harvard psychiatrist works with the left- and right-brain characters in an attempt to help his patients heal issues related to mental illness, the book *Of Two Minds* by Dr. Fredric Schiffer is a real eye-opener. It even addresses how our two hemisphere characters are so different that each may actually manifest unique aches and pains that the other does not acknowledge or exhibit.

Moreover, if you are looking for an alternative tool in how you might manage mental health issues, Dr. Richard Schwartz's Internal Family Systems model is an interesting strategy that recognizes and works with the different *parts* of a person's personality so they can collaborate to find a healthy solution. Each of these books and tools is fascinating if you are interested in learning more about the brain.

Both of our cerebral hemispheres are constantly contributing to the whole of any experiential moment, so I do not mean to imply that either the left brain or the right brain functions in isolation. Modern technology shows clearly that at any moment in time, both hemispheres are contributing to the input, experience, and output of the nervous system. However, as I stated previously, brain cells dominate and inhibit their counterpart cells as a standard practice, so the brain is not *all-on* or *all-off* under any circumstance except for death.

As we think about how the brain works, it is natural to ask this question: "How is it possible that a group of brain cells can work together to create a personality at all?" I'm not the first person to ask this question, nor am I the first to experience a brain trauma, exhibit a change in personality, and then recover the traumatized cells, regaining old circuits, old skill sets, and lost personality traits. However, I may be the first neuroanatomist to take this particular journey deep into the neural and psychological workings of my brain and then walk away with these unique insights concerning our Four Characters.

Brain cells are beautiful little creatures that come in many different shapes and sizes, and their design dictates their ability to perform their specific function. For example, the sensory neurons located in the primary auditory cortex of each of our hemispheres have a unique shape that supports their ability to process sound information. Other neurons that function to interconnect different regions in the brain have an appropriate shape for that action, as do the cells of the motor system.

Neuroanatomically, it is important to note that the neurons in your brain and how they connect with one another are essentially the same for all of us. Structurally speaking the bumps and grooves of the outermost cerebral cortex of everyone's brain are virtually identical, so much so that damage to a specific area in your brain would wipe out comparably the same function as in mine, should I have the same trauma. Using the example of the motor cortex, if you and I experienced damage to the same specific group of cells in the same hemisphere, we would more than likely experience paralysis in the same parts of our bodies.

Underlying the functional differences between our two hemispheres are neurons that process information in unique ways. For example, the neurons in our left brain function linearly: they take an idea, compare that idea to the next idea, and then compare the by-product of those ideas to the *next* idea. Therefore our left brain has the ability to think sequentially. For example, we know that we need to start the engine before we put the car in gear. Our left hemisphere is an amazing serial processor that not only creates abstract linearity—as in 1+1=2—but manifests for us temporality, the linear sense of time whereby we can separate past, present, and future.

Our right-brain cells are not at all designed to create linear order. Instead, our right hemisphere functions like a parallel processor, bringing in multiple streams of data that simultaneously reveal a single complex moment of experience. Our right brain manifests a rich composite of the right here, right now present moment by adding depth to the creation of our memories, which are influenced by both of our hemispheres.

Although many of our brain cells are responsible for doing obvious things like understanding language or manifesting vision, other neurons function to create our thoughts or emotions. We use the term *module* to describe the way in which groups of neurons interconnect with one another in order to function together as an aggregate. Each of our Four Characters, for example, is supported by its specific and unique module of neurons.

By the time my left brain finally shut down completely, I had drifted into the peaceful consciousness of my right brain, where I lost all sense of urgency.

When I experienced the hemorrhage in the left hemisphere of my brain, most of those cells simply went off-line because of the inflammation, swelling, and pressure buildup within my skull. In response to the trauma, the left-brain cells that had dominated my right-brain cells via the corpus callosum released their inhibition over those cells in my right brain, just as with the split-brain patients. When this occurred, the characters of my *left thinking* and *left emotional* networks receded, while the comparable characters of my right brain's *thinking* and *emotional* modules became untethered, unfettered, newly dominant, and free to run wild.

If you are curious about how I could recall the events that happened on the morning of the stroke, even with my left brain off-line, it is important to note that although the circuits in my left brain shut down due to the trauma, I did not die, nor did I become unconscious. In addition, the stroke did not blow up—boom—and then it was all over. Instead, after the blood vessel in my left hemisphere exploded, over the course of four hours, more and more blood slowly oozed into that left-brain tissue, shutting circuits off as it went along. I experienced the stroke as more of a slow pipe leak than an instantaneous power outage. As a result, my right brain retained the ability to replay the memory of the morning of the stroke, much like a video.

By the time my left brain finally shut down completely, I had drifted into the peaceful consciousness of my right brain, where I lost all sense of urgency. Temporally, my right brain existed solely in the present moment, with no past regrets, present fears, or future expectations. From this, and the next eight years of recovery, it was apparent that the job of my right-brain circuitry was to process the experience of the right here, right now present moment.

My left hemisphere, on the other hand, had functioned like a *bridge across time*: it was responsible for linking this present moment to the past moment and then to the next moment. Somehow those cells in my left brain were organized in such a way that I had been capable of linear thought. Miraculously, my left brain understood that I needed to put my socks on before my shoes.

Clearly, we have two hemispheres for a reason, and without our left brain we are completely nonfunctional in the external world, in that we have no past or future, no linear thought, no language, and no sense of the boundaries of where we begin or end. Our left brain offers us our individuality, while our right brain connects us with the consciousness of not only the collective whole of humanity but the vast expansive consciousness of the universe.

By having both of these hemispheres working together inside of one head, we experience a natural duality. As a result, it is normal for us to endure an ongoing internal conflict, based completely on the two uniquely autonomous perspectives of our left and right brains. For example, my left brain might want to jump on that homework immediately and get it done, while my right brain would rather go out and play, leaving the work for the last minute.

The Four Characters in Our Brain

The differences between the two hemispheres are far greater than simply the underlying anatomy, physiology, and resultant skill sets. My experience of losing my left brain and then rebuilding it over the course of eight years taught me that in addition to executing opposite abilities and constructing different realities, each of my two hemispheres is the home to very specific and predictable characters. These are the Four Characters you briefly met in the last chapter.

More specifically, with recovery, when I salvaged the functions of my left-brain *thinking* module (Character 1), with them came the goal-driven, well-organized, methodical, controlling character who had dominated my pre-stroke life. She was strong, powerful, capable, manipulative, good with time management, and completely judgmental. Following her recovery, she wanted to be the boss again inside my brain.

In addition, as this *Left Thinking* Character 1 regained its ability to process information linearly, as well as to judge things as right/wrong or good/bad, I recouped the ability to experience *emotions* that were dependent upon a memory from another place and time. For example, we have the ability to feel guilt or shame in response to something that has already occurred, or we can build up resentments over time, or seek revenge for something that has happened in the past. Once the emotional module of my left brain healed enough to come back online, I could experience these kinds of emotions again. But just as a strict and productive Character 1 came back online with my left-brain *thinking* tissue, a pained and cautious Character 2 came back online with my left-brain *emotional* cellular network.

I will admit that I truly enjoyed not having any of the pain from my past in my left-brain emotional module anymore, as I did not miss those tragic emotional circuits of my childhood. Yet having said that, life without the richness of deep emotions is a flat place to exist. This left-brain emotional Character 2 feels and knows our past pain, and it is this character that takes us right up to the edge of our potential growth and either pushes us over that edge or retreats us back into the safety of what feels familiar. I have to know what is safe and not safe if I am to define the boundaries of my safety. I have to know what is right for me in order to know what is not right. I have to know darkness or sadness to recognize light or joy.

It is our left-brain *emotional* Character 2 who screams, wails, and rages against all those injustices that it has perceived as hurtful, dangerous, or unfair. It also holds us back, flees or freezes when something triggers our fear. Over the years, it has been the job of this tender and vulnerable character to hold our past pain in memory for our future protection. If we want to evolve into our best selves and live our best lives, we must create a healthy relationship with our left-brain Character 2. We grow and thrive when we are brave enough to stand in the center of our own pain and listen to what it is trying to communicate.

During my recovery, the newly dominant *Right Thinking* Character 4, who felt open, expansive, kind, and as big as the universe, was not too keen on letting the recovered stress-driven Character 1 of my left thinking brain waltz right back in to dominate my consciousness. Although I have to say I was thrilled to have those neural networks back in action, so I could speak and understand when others spoke to me and know the boundaries of my body again, I preferred embodying the never-ending peaceful gratitude of my Character 4's open heart. That is why I consciously chose to remain right-brain dominant. And if I can choose which circuitry I want to run, so can you.

As you move through this book, you will learn a great deal more about each of your Four Characters, what they feel like inside of your body, and how you, too, can choose who and how you want to be in any moment.

CHAPTER 3

Our Brain's Team: The Four Characters

The unsung beauty of the split-brain experiments that I described in the last chapter is that they support with convincing neuroanatomical evidence the existence of the Four Characters. Surgically separating our two brain hemispheres has scientifically shown us that they are not simply two anatomically separate halves of a whole. Instead, the two halves of our brain house completely different character profiles that each exhibit unique wants, dreams, interests, and desires. (Just imagine what pearls of wisdom we might have gleaned if Gazzaniga had chosen to give a Myers-Briggs test to each of the hemispheres of those commissurotomy patients.)

For some reason that I am unsure of, modern science has back-pedaled away from many of the insights we gained from the split-brain research of the '70s, specifically concerning the diverse and often antagonistic characters residing in each half of our brain. Perhaps this idea faded simply as collateral damage when scientists rushed in to squelch the exaggerated public hype. Or perhaps not everyone, including the scientists involved, was cognizant of their own multiplicity of personality, and as a result those original seeds of knowledge did not receive the water they required for future growth.

In the "My Stroke of Insight" TED talk, I purposefully went out on a limb when I stated, "Our two cerebral hemispheres think about different things, they care about different things, and, dare I say, they have very different personalities." Be it a popular idea or not, I am bringing buckets of water to revive this very important conversation.

How Your Four Characters Think and Feel

Here is a brief list of some of the attributes exhibited by the Left Thinking Character 1 and Right Thinking Character 4 parts of our brain. Notice that these two thinking characters are virtually opposite in how they perceive and process information:

Left Thinking Character 1	Right Thinking Character 4
(Serial Processor)	(Parallel Processor)
Verbal	Nonverbal
Thinks in language	Thinks in pictures
Thinks linearly	Thinks experientially
Past/future based	Present moment–based
Analytical	Kinesthetic/body
Focuses on details	Looks holistically at the big picture
Seeks differences	Seeks similarities
Judgmental	Compassionate
Punctual	Lost in the flow of time
Individual	Collective
Concise/precise	Flexible/resilient
Fixed	Open to possibilities
Focus on ME	Focus on WE
Busy	Available
Conscious	Unconscious
Structure/order	Fluid/flow

Here is a brief list of some of the attributes exhibited by our Left Emotional Character 2 and Right Emotional Character 3 parts of our brain. Notice that these two emotional characters are also virtually opposite in how they feel when they experience emotion:

Left Emotional Character 2	Right Emotional Character 3
Constricted	Expansive
Rigid	Open
Cautious	Risk taking
Fear based	Fearless
Stern	Friendly
Loves conditionally	Loves unconditionally
Doubts	Trusts
Bullies	Supports
Righteous	Grateful
Manipulates	Goes with the flow
Tried and true	Creative/innovative
Independent	Collective
Selfish	Sharing
Critical	Kind
Superior/inferior	Equality
Right/wrong, good/bad	Contextual

In Part II of this book, we will take a much deeper dive into the skill sets and personalities of these Four Characters. I will not only help you identify each of these characters inside of yourself, but we will explore how your Four Characters can work together to become a healthy team inside of your brain. In Part III we will observe the Four Characters in action, or as I like to say, "in the wild." There, we will first take a look at how the Four Characters view their relationship with our body and then get a glimpse at how they each interact predictably in romantic relationships. Because it is our ultimate goal to create more connection and consequently greater health within ourselves and with others, we will take a peek at how devastating addiction can be to our Four Characters, and gain some insight into why it is that recovery may be effective for one person and not another. From there we will look at the evolution of the Four Characters over the last 100 years, and the profound impact that new technologies have had on the different generations.

As we go along in Part II, for clarity's sake I will share with you the name that I have chosen for each of my Four Characters, along with some of the things I know about her. I do this in an attempt to help you better relate to and identify that specific character inside of yourself. I believe it is vital that you take ownership of your own Four Characters, which is why I have chosen to not give each of these characters a generalized name other than Character 1, 2, 3, and 4. I think it is really important that you spend a little time contemplating a name for each of your Four Characters that is meaningful to you.

With naming your characters, please feel free to be as tame and proper or as absolutely outrageous as you want to be. Some folks have chosen names of their parents or friends, while others have selected mythological or fictional designations. Feel free to use a derivative of your own name or something completely off the wall. The point here is that you choose a name that will bring that character full force into the forefront of your mind when you refer to it.

Anatomically, each of us has a whole brain, and we each have all Four Characters. You may find, however, that one of your Four Characters may be dominating, or another part may rarely show up. If it turns out that you absolutely cannot identify with any of the Four Characters, you might ask your spouse, or a trusted friend, if they know that part of you. Please note that although we all exhibit thoughts, emotions, and behaviors that we are not proud of, none of our Four Characters is bad, wrong, or not worthy of our love and respect. In addition, it is not unusual that our perspective of ourselves is different from how others view us. Hopefully, whatever insights you gain will prove to be an important tool for your own personal growth.

Your Brain Team and Your Power to Choose

We have seen how these Four Characters are the natural by-product of our brain hemispheres' cells, circuits, and functional modules of thinking and emotional tissue, but what does this mean for you in your daily life? Just think about it. Does a day go by when you don't experience an intrapersonal conflict? Our two hemispheres value completely different things, so when your heart says one thing and your head says another, it is simply a dispute between the different parts of your brain. For example, a conflict between your right and left *thinking* characters might look something like this: "Do I take that job in a new city that pays more and is an obvious promotion (left-brain thinking values)?" or "Do I keep my current job so my children can stay in a familiar school and stay connected to their friends and family (right-brain thinking values)?"

Similarly, a conflict between your right and left *emotional* characters might sound like this: "That person hurt me so badly, I just want to get even and hurt them back (left-brain emotional values)" or "I'll just send that person love from afar, and create the time and space away from them that I need, so my heart can heal and I can move on with grace (right-brain emotional values)."

In each of these instances, knowing which characters are engaged in the dialogue, and what their motivating factors are, enables us to make conscious choices about who and how we want to be.

> **When your heart says one thing and your head says another, it is simply a dispute between the different parts of your brain.**

As you become proficient at identifying your Four Characters and learn to appreciate and value the skill set that each brings to your life, you will be able to make this choice more consciously and deliberately. However, just knowing your Four Characters is not enough. The ultimate goal is for your Four Characters to become so familiar with one another that they create healthy relationships among themselves. Once this happens, your Four Characters will collectively function as a healthy team that is armed with all of your genius and natural abilities.

Teammates in any situation—like a sports team on the field or a team of colleagues at work—will call a quick "huddle" to assess a situation and strategize their next moves. Your brain team, made up of your Four Characters, can huddle together at any moment to analyze what is going on in your life and then collectively decide who and how you want to be in the next circumstance.

Following up on that, in Part II not only will we explore the Four Characters in detail, but I will share with you a five-step process that I call the Brain Huddle, whereby we consciously take a pause, call all Four Characters into our awareness, and then together as a team contemplate our next best move. I will encourage you to practice the Brain Huddle in everyday moments so you can train your brain to make important decisions swiftly and skillfully. If you are willing to train your Four Characters to function as a team during the benign moments of your life, when you are in duress you will have that skill available.

For now, here is a quick preview of the steps we will take to do a Brain Huddle:

Breathe and focus on your breath. This enables you to hit the pause button, interrupt your emotional reactivity, and bring your mind to the present moment with a focus on yourself.

Recognize which of the Four Characters' circuitry you are running in the present moment.

Appreciate whichever character you find yourself exhibiting, and appreciate the fact that you have all Four Characters available to you at any moment.

Inquire within and invite all Four Characters into the huddle so they can collectively and consciously strategize your next move.

Navigate your new reality, with all Four Characters bringing their best game.

You will no doubt realize that these five steps of the Brain Huddle spell out "B-R-A-I-N." While I, of course, think this acronym is adorable, it also has a real purpose: to help you quickly remember the steps when the pressure is on and your Character 2's stress circuitry is running on overdrive. In moments like those, when you can barely think because the chemicals of anxiety or fear are flooding through your bloodstream and overwhelming your circuitry, this B-R-A-I-N acronym can beam like a bright neon light, reminding you of the steps you can take to call your brain team together so you can find your way back into the peace of your right brain.

This process of the Brain Huddle whereby we can consciously and deliberately bring all Four Characters into the conversation is both powerful and empowering. We have the ability to interrupt the automatic circuitry of our emotional reactivity and consciously choose which of the Four Characters we want to have as dominant in any moment. Knowing our Four Characters and being able to recognize them in others enables us to interact more authentically in a whole-brain way. We have the power to purposely build healthy and healing relationships with others.

Your Hero's Journey to Peace

As I noted in Chapter 1, this journey you are embarking on as you get to know your Four Characters and learn how to integrate them into your brain team is a mirror of the Hero's Journey in Joseph Campbell's classic monomyth. In addition, it is worth noting that the Four Characters coincide distinctly with Carl Jung's four major archetypes of the unconscious mind: Character 1—the Persona; Character 2—the Shadow; Character 3—the Animus/Anima; and Character 4—the True Self.

In the classic story of the Hero's Journey, the hero heeds the call to leave behind his rational, ego-based consciousness that processes the reality of the external world. In the language of the Four Characters, the hero must step out of the ego-based consciousness of his Character 1 left thinking brain to enter into the unconscious realm of his right brain. To embark upon this quest, the hero must be willing to let go of his possessions and worldly knowledge and embrace the death of his ego's individuality. To paraphrase Einstein, we must be willing to give up what we are, in order to become what we will be.

As you might imagine, this is an enormous task for the hero to undertake, which is why, of course, he is described as heroic. He must be willing to set aside everything he has acquired and grown up to be. (Much like the journey of the Buddha, who famously set aside his position and worldly possessions in order to grasp the true nature of reality and attain enlightenment.) But once the hero chooses to shed the rational, ego-based individuality of his left brain, he enters into the realm of his unconscious right brain, where he will meet the Anima/Animus, the androgynous nature of his soul. The hero cannot be both characters—his individual and collective selves—in the same moment. He must lay down the justice-demanding judgment of his dominating left brain (Characters 1 and 2) if he is to embody the merciful characters of his compassionate right brain (Characters 3 and 4).

When we are born, we have no sense of individuality, and our two brain hemispheres are similar in both their structure and in what they value. Over time, however, our left-brain cells develop the ability to define the physical boundaries of where we begin and end, and with that identification of self, we gain the ability to perceive ourselves as individuals who are separate from the whole. It is in those moments that the droplet of our left brain's individual consciousness becomes separate from that sea of cosmic consciousness from which it came. Before the hero's left-brain ego-cells developed his perception of himself as an individual, he possessed the collective knowledge of his right brain's unconscious mind. With time, as the individuation of his left brain developed, it grew to dominate and inhibit the knowledge of his right-brain mind. Consequently, the cosmic consciousness of his right brain shifted into the background, becoming his unconscious intuition.

It is said that in that moment when the hero lays down the sword of his left-brain righteousness and ego, he is emancipated from his left-brain individuality, dissolving back into the cosmic consciousness of the universe from which he originated. Like the droplet returning to the sea, the hero is instantly enveloped by the blissful euphoria of the eternal love that his soul once knew before he was born. Like the great whale that he had forgotten he was, his soul returns to gliding through the sea of silent euphoria at One with all that is.

Once the hero has battled his fear of death and all the other left-brain monsters he had been clinging to throughout his everyday life, he is now free to gain the insights of his heroic quest while enveloped in the wisdom of his euphoric right brain. At this point, however, the hero must choose to either return home and share his hard-won whole-brain knowledge or keep for himself the lessons he has gleaned. Returning home, he is different now, and it is his challenge to figure out how to live a balanced life in the external world while remaining aware of both his conscious and unconscious characters and their conflicting values.

I invite you to embark upon your own Hero's Journey as you explore the Four Characters inside your conscious and unconscious whole-brain hemispheres.

The Four Characters, as I outline them in this book, provide a neuroanatomical road map of the time-tested paradigm of Jung's Four Archetypes. Like a house with four rooms, two upstairs and two downstairs, our brain is the home of all of these Four Characters. With minimal effort, we can train ourselves to identify each of them within our own psyche, purposefully create healthy relationships between them, and then let them collectively, as our brain team, lead our lives in peace.

If you are willing to pause and recognize what is already going on inside of your brain, if you are game to observe how you present yourself under different circumstances, and if you are prepared to bring your present-moment awareness to your current thinking and emotional patterns, you will be well on your way to living a life of choice. I invite you to embark upon your own Hero's Journey as you explore the Four Characters inside your conscious and unconscious whole-brain hemispheres.

Peace really is just a thought away.

A Note to Your Four Characters

As you continue through this book, to each of your Four Characters I say this:

Left Thinking Character 1

My message to your Character 1:

Breathe. Be open. Exhale. I dare you to finish this book. Feel free to judge this material with caution, but please do it with an open mind. I know you are going to be focused on my typos or errors in semantics, but if you allow yourself to look beyond those details, you will be rewarded with tools you can use to create more order in your world and obtain a greater feeling of connection with those around you.

Titles your Character 1 might give this book:

Know Your Brain, Own Your Power

Control Your Brain: Live Your Best Life

Your Success Starts in Your Brain

The Why behind Emotional Intelligence

What your Character 1 might say about this book after reading it:

"Left brain, right brain, breathe."

"I'll be damned, those other parts of me actually have value."

Left Emotional Character 2

My message to your Character 2:

It's okay. You are not going to like this, and it's still okay. I hear you. You matter. You are the voice of alarm that protects us all, and as such you are an important part of the whole. This material will help the other characters better understand you, keep you safe, and value you. You are indispensable, as you are our growth edge. Without your guidance, we cannot stay safe and we cannot evolve into our best selves or live our best life.

Titles your Character 2 might give this book:

Feelings Matter

Your Feelings Are Valid

Master Your Pain

We Are Feeling Creatures Who Think

What your Character 2 might say about this book after reading it:

"It's okay for me to feel what I'm feeling."

"I can be happy. I can accept. I know why I feel the way I do. I matter. I am okay. I feel empowered. I am the key to living our best life."

Right Emotional Character 3

My message to your Character 3:

Of course this book is available on audio! You can stay in action and enjoy this book too. I know you would rather go do something really exciting right now, but if you are willing to grasp this material and incorporate it into your life, the other characters will recognize how important you are and give you more time for play and innovation.

Titles your Character 3 might give this book:

My Brain Is Super Cool

The We inside of Me Are Total Rock Stars!

Four Play

Our Brain: The Whole Enchilada

What your Character 3 might say about this book after reading it:

"Life is better than I even imagined."

"I love being connected to us all."

Right Thinking Character 4

My message to your Character 4:

Here lies the key for you to unlock all that keeps you small and contained in this life. You are our connection to our Higher Power, as you are clear that it is our number one job to love one another. Not just those outside of ourselves, but the various characters within. This material will help your left-brain characters find the balance between what they do and who they are. You are the peace that is just a thought away.

Titles your Character 4 might give this book:

Free to Be You

We Are the Life-Force Power

Befriend Your Brain

Peace Is Just a Thought Away

What your Character 4 might say about this book after reading it:

"We are One."

"Keep reading . . . the jelly is in the center of the donut for a reason."

Part II

YOUR FOUR CHARACTERS

CHAPTER 4

CHARACTER 1—
LEFT BRAIN THINKING

Our brain's left hemisphere is the primary tool we use to interact with the external world. On the morning of the stroke, those cells making up the network of my Character 1 left thinking brain were swimming in a pool of blood that rendered them completely nonfunctional. Besides losing a group of skill sets that were dependent on those brain cells to function, as I mentioned earlier, when my left thinking network went off-line, a specific part of my personality, a character that I had known for decades as my ego-self, also disappeared.

When my Character 1 left thinking cellular network was incapacitated, I could no longer identify the physical boundaries of where I began and where I ended. I have to say that even as a neuroanatomist, I had never been taught that there was a group of cells in my brain that did that. With those cells off-line, I perceived myself to be a gigantic ball of energy that blended fluidly with the rest of the energy in the universe. I felt so vast that I believed I would never be able to squeeze the enormousness of myself back inside this tiny little body. As you might imagine, one part of me found this shift in my awareness amazingly insightful and exciting, while my Character 1 would have judged this loss of my *self* to be degrading—if it had remained functional enough to ponder the idea.

In addition to not being able to perceive the boundaries of where I began and ended, my left brain could no longer ascertain the edges or boundaries between anything else in the external world. As a result, I experienced myself as fluid, in flow with the energy of everything en masse around me. This shift in perception was possible because our left brain is designed to perceive differences and separation at the level of things, rather than at the subatomic level of the particles that make up those things. The latter is the realm of what we call our unconscious mind, the territory of our right brain.

THE FOREST AND THE TREES

On the afternoon of the stroke, I learned that the energetic flow of all mass moves so slowly that it cannot be detected by our left brain. As long as the left brain is focused at the level of solid things and is preoccupied with detecting the details that allow us to differentiate one thing from another thing, it cannot focus on the component pixels that make up those things. In other words, our left brain focuses on the details that distinguish one thing from another thing (trees), while our right brain focuses on the pixels that have no distinguishing characteristics and move as one (forest), as a part of the cosmic flow.

Because our two hemispheres process information in these two opposite ways, our overall perception of the world is a blended combination of the big picture (right brain) and the details (left brain). Like an eagle soaring from a great height, it can perceive the enormous landscape below and still focus on that vulnerable (and delicious-looking) prairie dog half a mile away.

When my left brain went off-line and I could no longer detect information at the level of things, to use the example of the eagle, I lost the ability to distinguish the prairie dog from the landscape. I could only perceive the pixelated atoms that make up space and exist at the level of the cosmic flow. Consequently, when I stood in the shower on the morning of the stroke, I could not distinguish the pixels making up my arm from those composing the wall. All I could detect was my energy, blended with the energy making up the space around me. My perception of myself bypassed all boundaries, and I literally became as big as the universe.

When the language centers in the Character 1 left thinking part of my brain fell silent, I lost all ability to communicate with others and even with myself. Not only could I not speak or understand when others spoke to me, but I could not distinguish letters or numbers as symbols that had meaning. Pre-stroke, I had known who I was because there had been a group of cells in my left brain that manufactured my identity as Jill Bolte Taylor. These cells that made up my left-brain ego-center knew who I was, where I lived, and tons of other details like what my favorite color was. These ego-center cells had worked day in and day out to keep me abreast of all the tidbits, details, memories, and likes and dislikes that had made up my identity. I, Jill Bolte Taylor, existed because the cells in my left-brain ego-center told me I existed.

It's a bit disconcerting to think that who we are is completely manufactured by a small group of cells in our left brain . . . but that is how fragile our ego identities are.

When those cells of my left-brain ego-center shut down, and I shifted into the oblivion of my right brain, I had no idea who I was and I could not recall anything about my pre-stroke life. It was not as though I was missing a memory that I simply could not put my finger on; it was more like that memory (and I myself) had never existed at all. I know it's a bit disconcerting to think that who we are is completely manufactured by a small group of cells in our left brain, and that we can lose ourselves at any moment, but that is exactly how fragile our ego identities are.

Losing and Regaining My Character 1

In addition to losing all of those very important abilities and functions, when my left brain went off-line I also lost the characters of my left thinking and left emotional cellular networks. Like a stove with the two left burners turned off, most of the cells in my left brain were still physically there, but they were traumatized and could not function. Without the linearity of time that my left-brain cells had so effectively engineered, all I had was the vastness of the present moment. Unlike the hero in the Hero's Journey who willingly chose to set down the sword of his left-brain ego, mine was involuntarily stripped away. As I unwittingly shifted into the unconscious realm of my right brain, in the absence of a functional left brain I was rendered as inept as an infant.

The loveliest part of losing my left-brain emotional Character 2, which we will talk about in the next chapter, was the complete absence of anger and fear. Without any of the left-brain memories from my past overshadowing the present-moment experience of my right brain, I shifted into a state of blissful euphoria. Of course, as tantalizing as this experience was, the absence of my left-brain Character 1 rendered me literally half-witted, and I could not function in the real world. (Mind you, while I was in that condition of ineptitude, I did not feel any alarm about it.)

Over the course of eight years of recovery, as the circuits in my left brain regained their functions and became strong again, my left-brain characters eventually recovered and came back online. My Left Thinking Character 1, as I mentioned earlier, wanted to take over and be the boss inside my head again. But although she had been both an effective and brilliant part of my pre-stroke life and I had achieved a high level of success under her leadership, I was no longer motivated by the external factors of money and prestige that she valued.

Although I knew I would have to earn a living again, my right-brain characters valued a more peaceful life, at a slower pace, with more time spent sharing deep and meaningful connections with my family and friends. My mother had been caring for me post-stroke, and had just turned 70, while my dad was in his early 80s. Moving back to Indiana, where I could enjoy time with them while they were still alive, became my priority. I had learned how fragile life is, and the preciousness of true and meaningful connections took center stage in who I wanted to be.

Before the stroke, I had been proactive and willing to compromise my relationships by moving away from both my home and those I loved for the status and monetary gain of a career as a neuroanatomist at Harvard. Although I was truly grateful to recover the vast and important abilities of my left-brain Character 1, post-stroke I was no longer willing to choose the treadmill of work, work, work. Pre-stroke, my left-brain Character 1 had defined success through the achievement of external reward. Post-stroke, my right-brain characters found meaning through the internal standards of loving, being loved, and being in service to others.

I have named my recovered Left Thinking Character 1 "Helen" because she is hell on wheels and she gets stuff done. I have learned that I am completely dependent on Helen if I am to be a functional human being in the external world. However, as much as Helen would like to restore her crown and become the dominant character in my brain again, that's not going to happen.

Helen is, by any measure, a fantastic character, and I am truly grateful that she has come back online so I can be competent again. But Helen is neither my friendliest nor my *best* self, so much so that when my friends call on the phone and recognize Helen is in charge, they say "Hi, Helen," and lovingly ask if I will return their call later in the day.

LEFT-BRAIN CONSCIOUSNESS

Our left brain is designed to create order out of the randomness of the cosmic flow, and just as the eagle is capable of zooming its focus in on that prairie dog, our left brain is capable of distinguishing two items as separate from one another by examining their differences. Once our left brain identifies two things as separate, it can then organize and categorize those things, based upon their details.

For example, I can distinguish between a donkey and a boat because they are two very different things, sharing only a few similarities. With a little refinement, my left brain can tell a donkey from a monkey because, although they share similarities such as limbs and a head, there are still a lot of differences. With yet a higher level of refinement in my left brain's ability to process details with an exacting precision, I can differentiate between a donkey and a horse. Although structurally they look a lot alike, I can identify their subtle differences and categorize them appropriately.

Besides the ability to differentiate between things, somewhere along the line our left brain manages to manifest both an identity and a consciousness. Without writing a dissertation, for our purposes I will define consciousness for the left brain as an awareness of itself and its relationship with the external world.

The primary building blocks of the physical world are things, and as I noted above, our left brain is the masterful tool we use to perceive a thing as separate from the background of the cosmic flow. Our left brain identifies individual things by shifting its perception in such a way that it can compare, critically analyze, and distinguish minute differences in the structure and texture of particles in flow. By focusing completely at the level of things, our left brain creates a new level of consciousness.

You might remember back in the '90s when the Magic Eye stereo images became a fad. Two images were merged together into one image, and depending on how you focused your eyes, you could see either the most obvious two-dimensional image or the three-dimensional image that was hidden inside. The shift in our plane of focus that characterizes the left brain's perception is not exactly the same as what was going on with the Magic Eye images, but the principle is comparable.

In addition to viewing the external world at the level of things, through refinement and higher levels of differentiation our left brain defines the *edges* of where we begin and end as individuals. It does this by manufacturing a holographic image of ourselves so we can delineate between what is inside of us and what is outside of us. When this happens, our left brain perceives the physical world as separate and concludes there is an external reality and an internal reality.

The external world and our relationship with it move to the forefront of our focus because now we are separate from the whole. This means we are no longer safe because with life and separation from the cosmic flow comes the threat that there is now something we can lose, which is the "me-self" as the center of the universe, as well as life itself. Because we have become the center of our own universe, our left-brain ego-cells come online and begin organizing everything in the external world around our individuality.

With this shift of focus toward our relationship with the external world as separate from ourselves, the consciousness of the eternal flow remains, but shifts into the background. Focused on the prairie dog, our now conscious left brain ignores the background landscape, and the holistic realm of our right-brain perception is set aside.

With our left brain now manufacturing a new level of consciousness whereby we perceive things and our relationship with those things as outside of ourselves, we generate higher levels of order that eventually reach an advanced level of sophistication. The cells of our left-brain thinking Character 1 organize, categorize, count, list, and eventually name everything once they have structurally manifested language for communication with others.

As we saw in Part I, with the addition of our higher cortical thinking tissue, we humans have not only added new cells and circuits but we have achieved a functional consciousness that places us at the top of the food chain. By gaining the ability to think rationally, we have become able to create predictable routines and mechanically fit things together based upon their structure. These order-driven cells of our left brain account for both our reality-based consciousness and our elevated status.

Life is an ongoing event, and the more we learn, the more our left brain wants to learn.

By this point, our left-brain skill sets are well developed, and Character 1 storms in with the intention of ruling the roost. Our Character 1 is our power in the world and also the face we show: as we saw in Chapter 2, it corresponds to the archetype of the Persona, which Jung defined as "a kind of mask, designed . . . to make a definite impression upon others."[1] As our alpha self, our Character 1 will compete when it is challenged and fight for what it believes in. Using its ability to discriminate between this and that, our Character 1 will define what is right or wrong and what is good or bad. In this way, our left-brain thinking tissue establishes for us a worldview and belief system within which we will make our decisions and evolve our life.

---

[1] C. G. Jung, Two Essays on Analytical Psychology (London: Routledge, 1992), p. 192.

At the same time our left thinking brain is manipulating data in both linear and methodical formats, it is simultaneously laying down new neural connections in response to novel stimulation. Life is an ongoing event, and the more we learn, the more our left brain wants to learn. Neuroplasticity is the ability of our brain cells to rearrange which other neurons they are communicating with, and this underlies our ability to learn new material.

Because our brain is a product of both its nature and our nurture of it, we have the power to voluntarily change the cellular structure underlying our thoughts and feelings. For humanity this means that for the first time in the history of life on this planet, as far as we are aware, we have the power to direct our own evolution to a higher level of intercommunication.

Let's do it with a full understanding about the different parts of our brain and the power we have to use our thoughts to change the anatomical structure of the cells underlying those thoughts. We can do this through meditation and mindfulness, of course, and by using the Brain Huddle we can strengthen the relationships between our Four Characters, making that level of open communication the norm inside of our brain.

Character 1 in the World

Our left-brain Character 1 has the ability to be purposeful and intentional as well as thorough. By grouping things in repeatable and predictable routines, our left brain can construct a physical world that feels familiar, and thus, although we are separate from it, we can feel safe within it. Having gained an individual consciousness, our left brain becomes a true master at organizing things in space. We organize hierarchically when we judge one thing more important than another, we manage time such that we can be punctual, and when we make plans, we are organizing our behavior across time.

Character 1s wake up in the morning and see the day as something to be conquered. They are eager beavers who rise early, love routines, and thrive on crossing things off their lists. On the job,

Character 1s are effective leaders and good at managing people, places, and things. They laser focus their minds on details and are extremely productive. They are highly critical of their own performance and consistently compare themselves with others. Every day is a chance to sharpen their skills, and it is important to Character 1s to bring their most efficient selves forward.

True to their nature, Character 1s must create order in the space around them, and they value neatness because appearances matter. Everything a Character 1 does is deliberate because if something is worth doing, it is worth doing well. Time has value, so Character 1s are not only punctual but often arrive a few minutes early. You can be certain that the Character 1s who arrived before you noticed if you were late.

Character 1s value material goods, buy quality products, take care of their things, and are sure to give you the stink eye if you don't put the stapler back from where you took it. Our Left Thinking Character 1 is good at earning, organizing, and investing money and is excellent at both self-promotion and confrontation.

By design, Character 1s are gifted rational thinkers with a strong mastery of reality, so they reason their way to their best decision. Because they have taken the time to deliberate why they think what they think, they take responsibility for what they do. If there is a perfectionist inside of your brain, rest assured it is your Character 1.

Thank goodness our human brain has evolved to have the Character 1 skill set. Because of their organizational skills in government, academia, and business, we exist in an orderly society. As a result of their natural abilities, as a collective species we have a neuron on which we can hang an idea. Besides being great at fixing things, cleaning up messes, and running tight schedules all at the same time, our Character 1 respects authority, obeys rules, and consequently tends to keep us from doing really stupid things.

CHARACTER 1 AT WORK AND AT PLAY

Let's take a glimpse at how our Character 1 tends to show up in a couple of real-life scenarios. We will explore this character in action in much more depth in Part III, when we look at how the Four Characters operate in different areas of our lives. Please consider this a brief preview.

Note that as we observe our Character 1 in the wild, in some cases we will distinguish between our "Soft 1," which occurs when our Character 1 is absent of emotion, and our "Hard 1," which is generated in response to an alarm sounded by the deeper left-brain emotional tissue of our Character 2. As this implies, when played in isolation, our Soft 1 tends to be kind, thoughtful, relatively available, and a great team builder. Our Hard 1, on the other hand, is generated in response to an emotional upset of our Character 2. Consequently, our Hard 1 comes online feeling as though it is managing an emergency, because it is doing just that even if the emergency is only inside of its own mind.

Fortunately the majority of Character 1s that I interact with are Soft 1s, who are organized, competent, efficient, and also kind. Sometimes, if we are brought up in an environment where our emotional alarm is routinely triggered, our Character 1 gets trained to be a Hard 1. If you are willing to explore the motivators underlying your Character 1, you may find that you are a Soft 1. But if your Character 1 has grown out of the stress and anxiety circuitry of your Character 2, your Character 1 may be more of a commander in chief than a team leader.

Prior to the stroke, my left-brain emotional Character 2 existed in a high state of alarm, so my pre-stroke Helen was definitely a Hard 1. I remember how sitting in a boardroom would drive me out of my mind because everything took so long. And every time someone veered away from the agenda, I felt harsh judgment and literally felt physical pain in my body in the form of severe anxiety. My Character 2 was always revved up because of the trauma in my childhood, and being able to relax was not in my vocabulary. But after I had the stroke and that little Character 2 circuitry was wiped out, so was my sense of desperation and urgency. When Helen came back online post-stroke, she was Soft and much more pleasant to be around because she was no longer riding a full throttle alarm from my youth.

Character 1 on the Job

Character 1 leaders think linearly and start projects at the logical beginning. In order to understand the Character 1 as a leader in business, however, we must distinguish between the leadership styles of the Hard and Soft Character 1s. The Hard 1 will lead a group like a cattle driver who circles the herd and drives the working team with a prod from behind. The Hard 1 is the leader of the team, not a part of the team. The Soft 1, in contrast, leads a team like a shepherd who circulates among their flock, helping them as they go.

The Hard 1 leader is an influential critical thinker who leads their team based on vetted ideas and data. They value their team as a workforce rather than as a group of people. For the Hard 1 leader, there are real consequences for success and failure. They value reason and believe emotions don't belong in the workplace. The Hard 1 considers emotions to be a vulnerable weakness, in both themselves and in others. They take pride in their ability to laser focus on work and see themselves as higher on the food chain than those working below them. The Hard 1 reinforces this separation of value by never saying "me too" or showing their vulnerable underbelly.

The Hard 1 leader is the head of the pack, not interested in connecting with the worker beyond the task at hand. They have a great poker face in the middle of chaos and uncertainty, so the team is never really privy to the true status of the overall project. Because the Hard 1 authoritatively directs the project from above and keeps the big picture of the project from the team, workers are kept in a small state of mind. This helps the Hard 1 maintain authority, as it is virtually impossible for anyone on the team to knowledgably challenge the wisdom of their leader when they don't really know what is going on.

The Hard 1 has a preconceived expectation about what it wants to accomplish and is focused on the results rather than the steps the team has to take along the way. Hard 1 leadership functions linearly and piecemeal, so holistic solutions to fundamental problems may not be identified until it is too late. If the team does not have the bird's-eye view of what it is trying to accomplish, and there is no road map for a successful execution, the team cannot predict or compensate for either minor or significant pitfalls when they arise.

As I noted earlier, the Hard 1 leader is driven by a state of agitation that is brewed at the level of its left-brain emotional Character 2. The Hard 1's circuitry is triggered to protect its Character 2, which is caught in a repetitive loop of fear. If the Hard 1 pauses or fails, it won't be able to outrun the monster that is nipping at the heels of its Character 2. Therefore rest and risk are hard for the Hard 1 leader, who is habitually up at 4:15 A.M. to get the edge on its competitor.

Hard 1 leaders are highly self-critical. When they score a win, they take it personally, but standing alone at the top of the project, the Hard 1 leader will still feel isolated and empty. They cannot be content, kick back, and be at ease because with each victory comes yet another mountain to climb and another monster to escape. Lurking just beyond success is the ominous sense of *What's next?* And when a Hard 1 fails, they fail hard.

Hard 1 workers, like Hard 1 leaders, function well in a highly organized environment. They need to have a clear definition of what success looks like because they operate from a fear of failure. The Hard 1 worker focuses on the big win at the end rather than celebrating the little successes along the way.

You will get top performance from Hard 1 workers, but they do what they are asked to do and nothing more. They don't have the insight needed to truly think for themselves or bring genius or insight to a project. But then, Hard 1 leaders are not interested in having anyone else involved in the thinking, as they see others as a threat to their superior status. As a result, a Hard 1 leader tends to work well with Hard 1 workers, providing those workers are not eyeing the leadership position.

When it comes to change, Hard 1s are a tough crowd unless they can use that change to promote their own advantage. For example, they will embrace new technology, like a shift in software or hardware, if they believe it will help them get ahead of the competition. However, although they will demand that the change take place, the Hard 1 leader will not be helpful in the execution of the change. Hard 1 leaders will endure the inconvenience of change as long as the end result gives them an edge.

Similar to Hard 1 leaders, Soft 1 leaders are strong thinkers, but they are compassionate in their understanding that their team is made up of people who both think and feel. Soft 1s lead a team with empathy and assume errors come from a person's ignorance rather than from their incompetence. A Soft 1 leader operates from the conviction that everyone is doing their best work, which stems from their best thinking. Therefore, if a problem presents itself mid-project, the Soft 1 leader merely needs to tweak the team's thinking to help everyone get back on track. The Soft 1 leader hangs around and is a vital part of the team, not separate from them, and is not afraid to say, "me too." Workers feel like they are working *with* their Soft 1 boss rather than *for* that person.

Soft 1 leaders create a vision and provide a road map for their team, so each member understands their role in the project. The Soft 1 leader defines what success and failure look like right from the beginning. Because they do so, everyone on the team is clear about their job, and they can feel safe because they know what is expected of them.

The Soft 1 leader is hands-on, but not micromanaging. Under these conditions, workers feel supported and valued at every step of the way. That is why, under the management of the Soft 1 leader, workers take pride in their efforts and engage not only mechanically but emotionally. As a group they become invested in wanting their team to succeed to the end goal, and off they go.

The Soft 1 is driven by its need to do its best and to make the world a better place. As a leader, the Soft 1 is an active participant in the team's overall dynamics. They keep abreast of everything going on in all of the departments, and in this way they shepherd the team forward. When it comes to change, the Soft 1 will embrace something like new technology, trusting that it will contribute to the overall success of the team and ultimately the company.

When a Soft 1 makes a winning effort, it is a win for the team, and the Soft 1 keeps spirit true by celebrating the little wins along the way. The Soft 1 views complications as opportunities rather than as failures, and thus minimizes the risk to each individual team member. Consequently, each worker feels safe to take their next step forward, and together they collectively advance to correct the wrong and create another small win. With every success the Soft 1 says, "We win," in contrast to the Hard 1, who says, "I win."

Soft 1 leadership sets up a congenial working environment where everyone can thrive and bring their unique expertise forward. For example, an audit by the IRS can unnerve just about anyone simply because of the fear of the unknown and the potential financial repercussions. At the same time, an audit is an awesome opportunity to have a professional check the accuracy of your system, free of charge . . . which is all good, providing they don't find anything amiss.

But even if there is an error, an audit is an opportunity for an organization to hit the reset button for its financial accuracy. Although the Hard 1 would push through an audit, driven by the fears of their Character 2, which is being chased by yet another monster called the IRS, the Soft 1 would see the audit as an opportunity to explore with the team. Who has had experience with this sort of project, and how might they best prepare? The Soft 1 would then break down the responsibilities and ask everyone to set a timeline for their accountability. The Soft 1 leader capitalizes on the power of collaboration and shared ownership of the project.

The Soft 1 worker thrives in an environment that holds people accountable and on a timeline for their production, and then celebrates the little wins along the way. Under these circumstances workers know where they stand with a project, as well as where they stand with the boss, so they can feel safe, relax, and do their job. That being said, the Soft 1 worker tends to feel safe in an environment that manages change slowly, rather than a space that is constantly changing its needs and demands. Large systems like the government, major corporations, and academic institutions are all enormous ships that tend to change course slowly and pivot a few degrees at a time. Under the right circumstances, a Soft 1 worker will thrive on service, and even venture beyond their job description to serve the team.

Character 1 at the Beach

Character 1s will go to the beach with an agenda to relax, get some reading done, and enjoy the sunshine. They will bring a well-organized bag that is filled with all the lovely things they will need, including towels (and those cute little clasps to bind it to the chair), a drink holder, a book or Kindle, a phone, a cover-up, sunscreen, and their Warby Parker prescription sunglasses. They will arrive in brand-name sandals and expensive, stylish sportswear.

> **Character 1s naturally attend to details,**
> **so sitting next to the trash is not an option.**

Character 1s will build a little beach station, a functional workspace that is in perfect order. They will have checked the expiration date on their sunscreen before they arrived and set a schedule for how much time they will expose each of their sides to the sun, based on the time of day and the intensity of the rays. Character 1s will move their chair like a sundial in response to the movement of the sun for maximum exposure, and they will know exactly how much reading they can get done before it is time to turn over. Character 1s are aware of those around them and take appropriate precautions to protect their electronics and other possessions in response to the unconscious migration of the children who are playing behind them.

Character 1s naturally attend to details, so sitting next to the trash is not an option. The Character 1s' organized workstation could be here, home, or in a car on a road trip. They are well organized, take care of their things, and can bring along their little bag of goodies with them, no matter where they go. If they choose to play volleyball, Character 1s are competitive, so they will play to win.

When Character 1s hang out in a group, inclusion is important to them, so on the beach the group may be wearing matching visors and have the same beach toys but in different colors. Character 1s don't want to stand out, so they set a schedule that ensures they will eat together, play together, and go to the bathroom together. And if anyone goes wild and gets a henna tattoo, they will all follow suit.

When a Character 1 is at the beach, their thinking left brain automatically seeks differences, so their eyes will compare different kinds of shells and notice the details of the local birds. They may even carry a little guidebook about the local fish and flora. If they find something special that they would like to collect, like shark's teeth, they focus their eyes and comb the beach for just that. If they happen to see a dolphin, they will get excited because they will want to share that story and that experience with their pals.

A Snapshot of Your Character 1

- Organizes Everything: *Even my spice rack is alphabetized, and my stapler and scissors are put back where they belong.*

- Categorizes Well: *The clothes in my closet are grouped by season, and my garage drawers and shelves are labeled.*

- Is Naturally Mechanical: *I am good at assembling IKEA furniture and Christmas presents for the kids. I really appreciate a good set of instructions, and a clearly written manual thrills me.*

- Is Neat: *Appearances matter, so I check my clothes, hair, and makeup before I get out of my car or jump onto a Zoom call.*

- Plans Well: *I run a tight schedule and leave time for the unexpected so I can be on time. I pay attention to quantities so there is always a backup in the closet or pantry.*

- Respects Authority: *I know exactly where I fall on the ladder of hierarchy. I respect those above me and may dismiss or be in charge of those below me.*

- Critically Judges Right/Wrong, Good/Bad: *I strive to be moral, and it is important to me to be right.*

- Is Detail Based: *I am excellent with details, exact with numbers, and an overall perfectionist in everything I do.*

- Counts Everything: *Whether it is the number of stairs as I descend, the amount of money in my wallet, or someone's failures, I am keeping track.*

- Is Protective: *I divide people into we and they. I protect my we against their they. We are right and they are wrong. We are better than they are, and our needs are more important.*

Getting to Know Your Character 1

At this point I encourage you to take some notes about your own Character 1. These questions are designed to help you identify with this part of your brain. The more familiar we become with our Character 1, the easier it becomes for us to recognize this character when it is present and to become this character by choice.

If you prefer to skip these questions for now and continue reading, please feel free to do so. I realize this level of personal reflection requires time, focus, and courage. When you are ready, these pages will be an invaluable tool for you to identify this part of your consciousness.

1. Do you recognize your Character 1? Pause for a moment and imagine yourself doing Character 1 tasks whereby you are creating order. Picture yourself in your office, or planning an event, or organizing things in your home.

My Character 1 is the consummate professional part of my brain, which thrives in the challenge of juggling multiple projects at the same time. She loves catching up on bills, chasing details for a presentation, and even keeping my taxes well organized. But my Character 1, Helen, is in a hurry, highly focused, and not my most patient self. She demands competence and efficiency from herself as well as from others. When Helen steps into a room, she assesses quickly who she needs to speak with, influence, or be influenced by.

2. What does Character 1 feel like inside of your body? Do you feel relaxed or excited when you are busy tending to details? Do you stand differently, or does your voice change? Do you feel tension in your chest, gut, or maybe your jaw?

Although I have a strong Character 1, this is not my primary type, so when I am running the circuitry of my left thinking brain, it is somewhat of a discomfort for me physically. Because my Character 1, Helen, is a bit uptight, her poker "tells," if you will, include a recognizable furrow in my brow, and she insists on clenching my jaw. It's pretty easy for just about anyone to spot my Character 1 because she speaks with a certain tone that is decidedly more monotone or interrogative than my usual vocal inflection. In addition, she is relatively stern and preoccupied with getting things done so I can cross them off the list.

3. What if you don't recognize this part of yourself?

If you absolutely cannot identify with this character, that is okay. However, because each of these characters stems from the underlying brain circuitry, odds are good that you are wired for these skills. Neuroanatomically speaking, any neurological infarct or developmental disease that has the ability to wipe out cells or block our ability to access circuits could interrupt our ability to experience this character. This is, of course, exactly what happened to my Character 1 when I had my stroke. Fortunately Helen was recoverable, and she did come back online.

If you have not experienced a major brain trauma and you are still having a hard time identifying your Character 1, you might explore whether there was someone in your past who not only discouraged this part of you but perhaps even criticized, shamed, or belittled it. When we are children, we are completely vulnerable to the judgment of those around us. Both positive and negative comments from those we are dependent on hold the power to have a long-lasting impact on who we grow up to be. In our attempt to not only survive but thrive in the company of others, it is natural for us to shift our behavior to match the acceptable demands of those around us. If it is dangerous for me to express myself in a certain way, I will not do it.

Once you begin to notice how and when your Character 1 expresses its skill set in your life, pay attention to what it *feels* like inside your body to be that part of yourself. Your Character 1 might show up as bright and bold. It might be extroverted and naturally bulldoze others, or it might be shy and diligent, not wanting to attract any attention. There is no right or wrong here, there is simply an awareness that this part of you exists. The more awareness you bring to it, and the more appreciation and validation you offer it, the stronger that circuitry will grow. In the long run, the more familiar you become with how this character feels inside of your body, the more power you will have to either step into or out of this character by choice.

For more insight into your Character 1 skills, you might just ask yourself: *When do I assert my authority? When do I make decisions for myself or others? How do I organize my time, my food, or my clothing? What am I responsible for—perhaps a pet, or the grocery shopping? What motivates me to be punctual or to wear appropriate clothes? Is there order in my drawers or cabinets? How am I with money, and how do I nurture my friendships?*

If you still cannot identify your Left Thinking Character 1, or if that part of you feels either unacceptable or a cause of embarrassment, consider combing through your past to see if there was someone, perhaps a teacher, parent, sibling, or even a friend who did not value your opinion or approve of you using your voice. Was there someone who insisted on being the voice of authority over you, or who felt the need to control your finances? Did anyone insist on taking care of the details of your life because they thought you were inept, or did someone keep track of your failures and then keep reminding you of your incompetence? The dance we do with one another is not always a healthy one, and although our Character 1 abilities are invaluable for living a successful life, it is possible for one person's Character 1 to oppress another's Character 1 through either harsh judgment or coy manipulation.

If all of that fails, you might consider seeking input from someone else—maybe a friend, spouse, or colleague—about how they see this part of you. They may have a stronger sense of your Character 1 than you do, or they may know exactly the condition of your car and drawers and agree completely that this part of you is pretty much missing in action. It is true that some people thrive on chaos and somehow manage just fine without a strong Character 1 even in a world that values order.

4. Assuming you can identify your Left Thinking Character 1, how much of the time do you let this character run your life, and under what circumstances?

As I mentioned, I have the utmost respect for Helen and her skills because she is awesome at what she does, and she makes my life run smoothly. However, there is a different character inside of me that I prefer to have as my primary character most of the time.

Let me say right up front that no matter which of the Four Characters tends to be dominant in you, we are all unique and our diversities are our strengths. Your Left Thinking Character 1 might be the best lead for you, but personally, I tend to want to spend a lot more time playing than Helen is comfortable with. That is why my Four Characters have put Helen on a schedule that all of my characters have negotiated. You may be just the opposite from me and choose to schedule your playtime, with your work time as your default. Rest assured: there is no right or wrong here, just as long as all Four Characters have an equal voice and can agree. For inner peace, all four need to be heard, respected, and validated.

My Character 1, Helen, keeps my life in order by making lists, but she does not create all the lists on her own. Instead, she has invited each of my Four Characters to cooperate with her. When any of my characters is aware that something needs attention, they add it to the list. Consequently, all of my characters work in support of Helen to help her be the best she can be, and that strengthens her resolve simply because she feels valued. When all Four Characters execute the Brain Huddle, which we touched on in the last chapter and will explore in depth in Chapter 8, they become a unanimous and fortified voice, leaving no room for temptation or procrastination.

If Character 1 is your primary character, hopefully you really like this part of yourself and know it well. Character 1s live in the real world, and they thrive when permitted to control details and create order in our lives as well as the lives of those around us. Just a heads-up, however: it is our Character 1 who tends to run our stress circuit, so it will be important for our other characters to help us stay healthy and well balanced.

5. As you think about your Left Thinking Character 1, can you come up with an appropriate name for it?

Helen is my "hell on wheels" personality, and she totally rocks out the to-do list. Thanks to her, I know the boundaries of where I begin and end, and she is the home of my ego-self and identity. Because Helen is great with details, I know who I am, I remember my past, I learn from my mistakes, and I can find my way home.

How about you? What feels right for you to call your Character 1? What are three things you particularly like about this part of yourself?

6. Who are some of the Character 1s in other people over the course of your life who have influenced you, in positive or negative ways? Was your Character 1 emboldened by their Character 1, or repressed?

I'm guessing it will be easy for you to identify some of the influential Character 1s in your life. When I was a child, my mother, G. G., was nicknamed "the hostess with the mostest" by her peers. She not only ran our home like clockwork but she shouldered her college classroom and my father's congregation of some 300 families. As a Character 1 overachiever, G. G. took the prize.

Having order in my life as a child was very helpful, but I can't say I always liked it. If G. G. was committed to anything, it was training her children to have strong Character 1s. This proved to be an uphill battle since, as it turned out, my father didn't have an ounce of Character 1 in him. However, with G. G. modeling a strong Character 1, and Hal modeling chaos, this combination really helped me value the gift my mother was trying to instill in me. Order is a beautiful thing, and it does make the world go around more smoothly.

Another really powerful Character 1 in my life was Mrs. Valerie O'Rear, my English: Advanced Composition teacher in high school. That woman absolutely terrified me, but for some reason, I performed better for her than any instructor before or since. She was both a big thinker and a detail monger who oozed with expertise. She gave us her best and expected the same in return. Even though she petrified me, she got my attention and I learned from her. Just look at me now . . . I write. Even though she is probably up in heaven moaning about my dangling participles and all those sentences I end with a preposition.

At the same time I can think of several insensitive Character 1s whom I probably could have lived without, even though they may have imparted valuable life lessons to me. When I first arrived at Harvard, most of my colleagues found my bouncy, friendly, and Midwestern enthusiasm refreshing. However, one of my Character 1 superiors made it perfectly clear that I was way too happy to ever be a serious scientist. I will admit that his unconstructive influence on my career may have been part of the reason why I was so motivated to win the department's most prestigious research competition. Although he may have forgotten our exchange, it made it doubly satisfying for me to receive the award. As a side note, it really is important that Character 1s be conscious of how their hostile judgment might negatively influence others.

How about you? Who were some of the powerful Character 1s in your life, and how did your Character 1, or other parts of you, respond to them?

7. Who in your life appreciates, cares for, identifies with, and wants to hang out with your Character 1? What are those relationships like?

Character 1s tend to prefer the company of other Character 1s, both professionally and socially. They care about the same sorts of things and really value their like-minded peers. In the same mode, I have learned that Helen is a unique duck, and not just anyone wants to be around her. Some of my friends who are strong Character 1s, however, are happy to team up and work on projects together, because there is nothing more productive than a committee of Character 1s who like and respect one another.

Consequently, Helen resonates well with my accountants, banker, and administrative assistant, but because Helen is a tool that I use to get my work done, when I'm not on the clock she fades nicely into the background. If you happen to be a primary Character 1, you and I are probably living very different lives. Having said that, thank you for making the world go around.

8. Who in your life does not get along with your Character 1?

When my father was 80, he flipped and spiraled a beautiful Miata convertible while traveling across the country. On that fateful day, I became his primary caregiver for the next 16 years. Prior to this accident, Hal and I were blessed with a fantastic relationship. We were friends, and we had a lot in common because we were both primarily energetic and creative Character 3s. But everything changed between us on the day of his accident. Instead of being his daughter and his pal, I became the authoritative female in his life, who looked like his mother and sounded like his ex-wife (G. G.).

Because of the accident, I had to step in to take care of Hal's financial world, his medical needs, and everything else that lands on a caregiver. It wasn't a job I asked for, but it was a job that fell to me as the only local child capable of doing it. The hardest part for me was that although my Helen had all the responsibility, she did not have enough actual power to protect Hal from being taken advantage of. In the big picture, Hal became resentful that he was now under my thumb, and he railed against the authority of my Character 1 when all I wanted to do was keep him safe. I'm sure he didn't realize that the more he rebelled, the more rigid my Helen had to become to keep the order. It was a very unpleasant experience for both of us.

I'm guessing your Character 1 has also had some pretty challenging relationships, with family members, friends, or perhaps even a colleague at work. It is often hard for others to voice their appreciation for the way our Character 1s work on their behalf, and isn't it refreshing when someone actually does express their gratitude? It is true that there may be a fine line between my doing you a favor and overstepping a boundary. I think it is important to remember that in general, Character 1s really are just trying to be helpful.

9. What kind of parent, partner, or friend is your Character 1?

Years ago I introduced a friend to her Four Characters, and she realized that she parented her two children completely differently. She showed up for her son as her creative Character 3, as she was his head cheerleader, and only offered advice when he solicited it. For her daughter, however, she parented from her Character 1, offering her opinion and consistently playing the devil's advocate. My friend noted that her relationship with her son was healthy and carefree, but her relationship with her daughter was stressed and often contentious. She promptly chose to shift her parenting style with her daughter to include her other characters, and their relationship immediately improved.

As we take a closer look at our Left Thinking Character 1, it is easy to note that this part of ourselves may appear as a bit cold, robotic, or emotionally unavailable. That is because this part of our brain is specifically designed to create order out of the disorder in the world around us. In its most pure form, our Character 1 is not designed to express emotions. Instead, as we discussed in Part 1, our left-brain thinking tissue Character 1 has been added on top of the left-brain emotional tissue specifically to refine and temper our potentially disgruntled Character 2. As a result, our Character 1 can often be found parenting, supporting, or even disciplining our own Character 2.

10. How kind is the relationship inside your head between your Character 1 and your other characters?

I may be jumping a little ahead of myself because we have not yet fully examined all Four Characters in detail, but my guess is that you are probably gaining a pretty good sense of how they each show up in your life. I am a true believer that the most important relationships we have are the ones that are being carried out inside of our own head. As a result, it is really important that we consider how your Character 1 views and interacts with your other characters.

When I was in early elementary school, I was sent home for wearing a striped shirt with plaid pants. Apparently the other little girls' Character 1s were offended by my mismatch, and the teacher thought it would be wise for me to change because they were belittling me. My little Character 3 didn't understand what the problem was because, from my perspective, I was happy wearing my favorite top with my favorite pants. My little intuitive right brain was not even capable of considering that this could be either a sensory violation or remotely problematic.

It wasn't until my college years that Helen came online full force. This makes sense because I was away from home for the first time, and G. G. was no longer convenient to keep the order in my life. Frankly, if I did not want to live like a total slob, I had to show up for myself and start taking care of things. I will admit that it didn't happen naturally for me until I fell in love with the study of anatomy, and academic success demanded a high level of order.

Predictably, the more organized and structured I became, the more I studied and the better grades I made. To my dismay, however, my left-brain Character 1 suddenly judged my fencing and tennis playing as a complete waste of time. Helen came on strong and found my more relaxed characters as unruly or disagreeable. At some unique time for each of us, we must learn to graciously negotiate the division of time we spend between work and play. I know that for many of us, finding this balance is one of our greatest ongoing personal challenges.

Cheat Sheet: Get to Know Your Character 1

1. Do you recognize your Character 1? Pause for a moment and imagine yourself doing Character 1 tasks whereby you are creating order. Picture yourself in your office, or planning an event, or organizing things in your home.

   _____

   _____

2. What does Character 1 feel like inside of your body? Do you feel relaxed or excited when you are busy tending to details? Do you stand differently, or does your voice change? Do you feel tension in your chest, gut, or maybe in your jaw?

   _____

   _____

3. What if you don't recognize this part of yourself?

_____

_____

4. Assuming you can identify your Left Thinking Character 1, how much of the time do you let this character run your life, and under what circumstances?

_____

_____

5. As you think about your Left Thinking Character 1, can you come up with an appropriate name for it?

_____

_____

6. Who are some of the Character 1s in other people over the course of your life who have influenced you, in positive or negative ways? Was your Character 1 emboldened by their Character 1, or repressed?

_____

_____

7. Who in your life appreciates, cares for, identifies with, and wants to hang out with your Character 1? What are those relationships like?

_____

_____

8. Who in your life does not get along with your Character 1?

_____

_____

9. What kind of parent, partner, or friend is your Character 1?

_____

_____

10. How kind is the relationship inside your head between your Character 1 and your other characters?

_____

_____

CHAPTER 5

Character 2—Left Brain Emotional

On the morning of the stroke, the cells in my left-brain Characters 1 and 2 experienced severe trauma and went completely off-line. Seventeen days following the stroke, the surgeons went in to remove a blood clot that was the size of a golf ball. That hemorrhaged blood had been pressing against my left-brain tissue, displacing the cells as well as blocking their communication.

Although some of the cells in my left-brain thinking Character 1 that knew how to calculate mathematics died from the trauma, within a month following surgery, many of the other Character 1 cells began to recover and communicate again. It would take eight years for all of my old Character 1 files to come fully back online, but in the case of mathematics, although I did get some skills back, I am no longer performing sophisticated proofs and equations. My left-brain emotional Character 2, on the other hand, was wiped clean like a motherboard and never recovered. Consequently, the trauma of the stroke forever stripped me of my emotional past.

As mentioned in Part 1, when we recall really fantastic or traumatic experiences, it is common for our thinking memory to be linked to our emotional memory. For example, when I was four years old, John F. Kennedy died. I remember that I was playing at a neighbor's house, and they sent me home when the news of the assassination first broke. I was too young to understand what a president or an assassination was, but I remember walking into my family home and feeling an odd sense of doom. I'm not sure if I had ever seen my mom cry before, and I remember feeling confused and scared.

Pre-stroke, when I recalled that day, I could replay the cognitive memory mixed with those feelings of doom. Post-stroke, however, when my Character 1 recalls the thinking memory of that day, my emotional Character 2 has no emotional content to contribute to the memory. Although I can remember that I felt a certain way, I cannot recapture the feeling anymore. To give you another example, one of the most important days of my life was when I graduated with my doctorate degree. Although I remember feeling tremendous pride that day, my recollection of that event is emotionally void.

As I noted earlier, the primary distinction between the brain of the human versus other mammals is the addition of the thinking tissue of both our left-brain Character 1 and our right-brain Character 4. As we have discussed, each of these thinking characters came online to directly modify and refine the underlying emotional limbic tissue of our left-brain Character 2 and our right-brain Character 3.

However, in order to truly understand who our emotional characters are, we need to understand that they have evolved to refine and modify the underlying structures of our *reptilian* brain. In fact, the neuroanatomical difference between a reptile and a mammal is the addition of this emotional tissue, just as the difference between the human and other mammals is the addition of our Character 1 and Character 4 modules of thinking tissue.

OUR REPTILIAN BRAIN

When we examine the functions of our reptilian brain stem structures, we can be grateful for their high level of automation. Thank goodness we do not have to tell our heart to beat or encourage it to race faster in the face of danger. Consider how exhausting it would be if we had to consciously remind ourselves to breathe. Our reptilian inheritance specializes in these fundamental activities, as well as regulating our body temperature, managing our balance, and driving our need to mate.

From a psychological perspective, our reptilian structures are all about our instinctual survival, and many of these circuits operate like on/off switches. Because these functions are necessary for our survival, the reptilian part of our brain is both rigid and compulsive, in that once a circuit is turned on, it will stay on until it is either satiated or exhausted. For example, I feel hunger until it is satisfied. I am thirsty until my thirst is satiated. It is fascinating to me that this part of my brain has to tell me to stop drinking water or I will drink myself to death.

Thanks to my brain stem cells, I am not tired until my brain shuts down my alert response and floods me with neurochemicals that make me sleepy. When I wake up, it is because I have an amazing group of cells specifically designed to arouse me. If something should happen to those cells and they stopped functioning, I might sleep my life away as if I were in a coma, because, well, I would be in a coma.

At this most fundamental level of information processing, I breathe because there is a group of cells in my reptilian brain stem that tell my diaphragmatic muscle of respiration to contract, and in response to that pull, air is sucked into my lungs. If those brain stem cells should be destroyed, I would die unless I was placed on a ventilator to help me breathe.

The cells at the level of our brain stem respond to incoming stimulation by triggering fixed patterns of behavior. At the same time, certain groups of these brain stem cells will determine whether we are attracted toward something or repelled by it. Think about your response the last time you felt something crawling on your skin. Your reflexive responses were a by-product of your reptilian brain and spinal cord connections that acted to swipe that creepy crawler off your body. Then almost immediately you probably felt an emotion like disdain flash through you. This is the ongoing one-two punch of these two different groups of circuits: the unconscious, automatic action of the brain stem immediately followed by an emotion that later infiltrates our consciousness.

The spinal cord is a well-organized structure that functions like a multilane highway for the transportation of very specific forms of sensation up to the brain stem, and motor action down from the brain stem. Different lanes carry unique forms of sensation up from the body to the complex area of the brain stem for processing. Some of the unique sensations include our fast-track pain fibers that transport acute pain, like one might feel in response to a predator's bite. As the name implies, intense pain will fly up from the site of the wound into the brain stem at record speed and automatically trigger a constellation of predictable responses that include vocalization (screaming bloody murder) along with a fight response for counterattack or push away. In comparison, our slow-track pain fibers carry dull or aching pain (like the type we feel with a chronic muscle problem) up to a different cluster of brain stem centers, triggering appropriate responses like stretching or reaching for a pain medication.

FEELING CREATURES WHO THINK

A lot of neuroscience research has been dedicated to mapping the connections between the cells in the brain stem and higher brain structures. At the same time, it has been challenging for scientists to tease out the whole story of what exactly is going on at this juncture between our brain stem cells and the newly added-on emotional tissue of our mammalian limbic system. Although we are quite clear on the function of many of the brain stem cell groups, because of the sheer density of fibers in this area, combined with the limitations we have in performing *tracking* research on living humans, some connections remain a mystery.

At a most basic level, it is the job of our amygdalae to conceptually ask moment by moment, "Am I safe?"

We do know that when the brain stem transfers its well-organized data up to the emotional tissue of our Characters 2 and 3, it is their job to modify and refine that data by streaming it through the filter of emotion. As we are feeling creatures who think, rather than thinking creatures who feel, our left-brain emotional Character 2 will eventually transfer much of its information up to our Left Thinking Character 1, while our right-brain emotional Character 3 will send its information up to our Right Thinking Character 4. Our two thinking brains will then regulate those emotional characters, as well as share their unique temporal modes of consciousness.

The limbic emotional cells positioned in our two hemispheres receive direct inputs from our reptilian anatomy. However, although our left and right emotional brains receive the same information from our brain stem cells, they process it in very different ways. To put it as simply as possible, our brain stem (including the midline midbrain region) sends information directly to the emotional cells of our two amygdalae. We have one amygdala in each of our two cerebral hemispheres, and it is their job to make an automatic threat assessment, based on how something feels.

At a most basic level, it is the job of our amygdalae to conceptually ask moment by moment, "Am I safe?" This safety may be either physical or emotional. The left amygdala tissue of our Character 2 module receives the information about the present moment and then immediately compares that information with our past experiences. For instance, let's say that when I was young, I had a bad experience with a bully who was tall and skinny, had blond hair, and wore a red ball cap. If at some point in the future I should run across another person who resembles that description, it would be the job of my left-brain emotional Character 2 amygdala to recognize those traits and sound my internal alarm.

Our right amygdala, on the other hand, does not compare our present-moment experience to our past. Instead, as we will see in the next chapter, it is completely focused on the richness of the experience of the right here, right now, present moment in all of its glory. Because our two emotional systems simultaneously assess our external level of threat in opposite ways, we reap the combined benefits of both: the big picture of the right here, right now as well as our wisdom from past experience.

It is critical to recognize that this divergence in how our two emotional brains assess threat gave rise to the duality of our two unique forms of consciousness. Our right-brain emotional Character 3 would dwell consistently in the conscious realm of the present moment and always perceive itself in direct relationship with the cosmic consciousness of the universe from which it came. But as soon as the consciousness of our left-brain emotional Character 2 shifted into the temporal realm of the past, it defined itself as an individual in the three-dimensional external world, no longer caught up in the flow of the whole but instead rendering itself as isolated and alone.

Henceforth, our right and left brains would forever evolve on separate yet parallel tracks of consciousness and exist in duality. Our right brain would evolve to be the home of our feminine, yin, and grace of the cosmic right here, right now, while our left brain would develop masculine, yang, and ego-centered traits based on our individuality and past experiences.

In general, when stimulation streams into our two amygdalae from our brain stem cells and enough of that information feels familiar, we feel safe and calm. However, as soon as one of our amygdalae perceives something as threatening, our danger alarm is triggered and our fight, flee, or freeze response is engaged. Because our two emotional systems are fundamentally different in how they process information and what they value, our emotional Characters 2 and 3 are equally different not only in what they perceive to be a threat, but also in how they automatically react. As a result, our two emotional brains are uniquely wired, and these differences have the potential to register as internal emotional conflicts.

Consider what is going on in your brain right now. Look up and look around, and then ask yourself: How does it feel to be in the space you are in? Is it an inviting place where you can feel comfortable, relax, and nurture yourself, or are you fighting a gnawing urge to clean up that mess over there so you can concentrate? Our two ways of feeling in the world are in action during every moment, and we tend to bounce quickly back and forth between them. Are we experiencing an emotion about something related to the past, or are we experiencing an emotion that is a response to the present moment and not influenced by the past?

ANXIETY AND FEAR

Physiologically speaking, just as our reptilian brain processes pain, aggression, pleasure, and our drive to mate, our two emotional brains are each committed to our self-preservation. They both seek to regulate our body's response to emotional stimulation and activate our fight, flight, or freeze automatic response, as they deem appropriate. The cells of our emotional system have the ability to increase our heart rate when we are nervous, afraid, or excited, as well as influence both the rate and depth of our breath.

In addition, we are completely dependent on our emotional tissue, specifically the cells of our hippocampi (one in each hemisphere), to create memories. It is important to note that our amygdalae have an antagonistic relationship with our hippocampi, in that when our amygdalae sound the alarm, our hippocampi shut down and we are no longer capable of learning and memorizing new information because we are too busy managing an emergency. Imagine how impossible it would be for a child who is living with high stress (amygdala alert) to try to learn anything at all when her learning brain has neuroanatomically shut down.

At a fundamental level, our emotional brains speak volumes to us through their expression of fear and anxiety, which run on different circuits and are generally triggered by different types of events. Fear is a strong emotion that is most often triggered in the present moment (right brain), in response to a known, definite, and immediate threat. For example, if I am walking through the woods and almost step on a snake that is slithering across the path, because I am terrified of snakes, my fear response will be instantly triggered. When this happens, I feel an intense flush of freak-out, I squeal like a pig, my heart races a million miles an hour, and I jump backward because blood has instantly rushed into my flailing arms and legs. My eyes dilate as I frantically look about to see where that petrifying thing just went. And, oh yes, to my complete embarrassment, I had forgotten that I was chatting on the phone with my pal, who probably heard the whole event and is now either laughing or terrified for me. All depending on which character in her brain is more dominant.

Although we also feel anxiety in the present moment, it is generally triggered by an experience or trauma that has already happened in the past (left brain) or we are anticipating that it will happen at some unknown time in the future. Anxiety feels like an overall body agitation that is accompanied by a feeling of despair or self-doubt. It is often triggered by a worry, fret, or concern that something unpredictable, unpleasant, or dangerous is lurking around, and thus we feel physically or emotionally vulnerable. Back to the example of that snake encounter, once the chemicals of fear have flushed through me and then dissipated from my bloodstream (90 Second Rule), I now run my anxiety circuit with the worry that I may come face-to-face with another snake, and I cannot shake this feeling of imminent peril.

It is so important to note that although we may be able to train our rational thinking left-brain Character 1 to override an automatic fear response, at the level of our neurocircuitry, we are feeling creatures who think. Denying our feelings can be detrimental to our overall health, and pent-up emotions tend to fester and stimulate our left-brain stress response, making it impossible for us to relax and find peace.

Although a rational cognitive left-brain Character 1 that values its self-control is a beautiful thing, when we train ourselves to ignore our emotions, or disregard what we feel as not valid, like a drainpipe that can get clogged up, those emotions will seep out one way or another. When the emotional pain of our Character 2 is not listened to or validated, it has the power to manifest as physical disease. As a result, it is our emotional Character 2 that often holds the key to our physical and mental well-being.

THE KEY TO OUR PEACE

Getting to know our Character 2 and learning how to nurture it with all of our characters in the Brain Huddle is how we will become healthy. I like to think about our left-brain emotional Character 2 as our superhero because it was so mighty that it was willing to shift out of the known, away from the familiar, away from its connection with God, the Infinite Being, the cosmic consciousness—whatever you are comfortable calling it—to exist in a whole new realm of consciousness as an isolated individual. Our Character 2 sacrificed its own peace of heart so we could evolve.

This willingness of our Character 2 sits at the core of our ability to process information linearly. Because we can break time down temporally into a past, present, and future, we have gained a new level of consciousness with the capacity to manifest order in the external world, which our Character 1 has mastered and refined to an artful extreme. When our Character 2 shifted out of the peaceful bliss of the present moment, it landed face-to-face with all the threats and monsters in our external reality. Including our potential for death, pain, or sickness, which lurk in our consciousness as an unpleasant possibility in every next moment.

It is this audacious Character 2 who faces our deepest fears and sounds our alarm for danger in the only way it knows how. It wails, it whines, it cheats, it schemes, it self-loathes, it gets jealous, it gets angry, it feels guilt and shame, and it acts out in a million antisocial ways to get our attention. Our left- and right-brain emotional Characters 2 and 3 can throw a temper tantrum at any age because the cellular networks of our emotional system never mature.

When our Character 2 shifted out of the peaceful bliss of the present moment, it landed face-to-face with all the threats and monsters in our external reality.

In addition, the cell bodies of our emotional brains are already in position and relatively well hooked up in circuits by the time we are born. As a result, we are wired to express ourselves emotionally as soon as we arrive in the external world. This is not true for our thinking cells. Although the cell bodies of our thinking characters have migrated into their position in the six-layer cortex by the time we are born, it will take years for those cells to interconnect. This is why it is so important that we purposefully stimulate the brains of our children with an enriched environment, early on.

The primary job of our emotional Character 2 is to filter out immediate danger and simultaneously help us focus our attention. These cells act by making comparisons and then attracting us toward things we want and repelling us away from things we don't want. At a cellular level, the power of our human brain is in our ability to inhibit automated circuitry and discern which circuits we want to run, versus which circuits we view as a distraction.

Say, for example, that we have a million thoughts and emotions bouncing around inside of our brain. Because our Character 2's power resides in its ability to inhibit and instinctively shut information out, it automatically narrows our focus. Combine this natural instinct and ability to push things away and say no with a predominantly discontent character who sacrificed its connection with the blissful cosmic consciousness of the universe, and you end up with a character who, at its core, is the source of our default *oy vey*. For many of us, we will spend the rest of our life trying to rediscover that feeling of eternal connection that we lost when our Character 2 opted to live in the illusion of the external world.

Living life with a suspicious and discontented Character 2 is one of the prices we pay to have a consciousness that is capable of focusing on the external world. At the same time, however, our Character 2 is also the source of our deepest and most profound emotions. Among others, we have the capacity to feel overwhelming loneliness, become completely enveloped by grief, and love more deeply than we could ever imagine. When we hurt, or hate, or feel completely possessed by jealousy or rage, these emotional experiences are both powerful and delicious.

I always share that I don't mind if someone is miserable as long as they remember to enjoy it. The ability of our brain to manufacture an experience of misery is a total art. We all hurt, and having the emotional capacity to experience true suffering is an amazing part of being alive. We just get into trouble when we spend too much time running that circuitry, believing that it is *our truth*, rather than realizing that it is simply a group of cells running a circuit. I feel pain. I am not the pain.

There is a mountain of neuroscience research that supports the idea that our left brain is the source of our happiness, and I agree with this completely. I do want to point out, however, that happiness is not the same as joy. Although both happiness and joy are positive emotions, they are very different psychologically and neuroanatomically. As many have observed, joy is cultivated from within. It comes when we make peace with who we are, why we are, and how we are, whereas happiness is contingent on external things, people, places, thoughts, and events. Because our experience of happiness is dependent on external circumstances, our Character 2 is the natural underlying circuitry of that happiness, or lack thereof. Our source of true joy is our right-brain Character 3, which we will expound upon in the next chapter.

For many of us, when our Character 2 is negatively triggered, because this is a part of our stress circuitry it tends to feel like a bolt of anxiety, fear, or emotional pain. Instead of our being in control of this character and choosing to run it on purpose, this character is apt to leap right in and almost violently take over our consciousness. Once Character 2 has stormed into our psyche, getting into another character requires a plan. This is clearly the perfect time for our Four Characters to take a Brain Huddle. Learning to master getting this character supported and instantaneously under control is the key to successfully managing our emotional reactivity.

Who among us has not secretly wished that we could just pluck this Character 2 part of our brain right out so we could escape our emotional pain from the past? Who among us has not sought therapy or guidance from professionals to help us fix, manage, or psychoanalyze why we are the way we are and why we feel the way we feel? And, most importantly, what might we do to heal it? The billion-dollar question, supported by a multibillion-dollar industry, is this: What are the strategies we can use to save ourselves when we realize we have devolved into our Character 2's emotional reactivity?

In the absence of my own left-brain Character 2, which had anchored me in the external world at an emotional level, my ego-self disappeared and with it all of the personal content of my identity. Consequently, I no longer existed as an individual who was separate from the flow of the universe, and I knew nothing about my life. Interestingly, my mother, G. G., even lost her "mother power," because not only did I not know who she was but I didn't know what a mother was. With no language available for me to label anything, or any individuation of things in the external world, it was impossible for me to think abstractly. By anyone's definition, I had become an infant in a woman's body.

Following brain surgery I did regain the ability to experience new emotions, but I had to learn all over again how to label the feelings I was experiencing. I remember describing the elements of feeling a pain in my chest. My heart was racing, there was a cramping in my jaw, the hair prickled at the back of my neck, my hands were gripped into fists, and sweat was pouring out of me. I felt like a wild, pacing animal, and I wanted to strike, bite, and go on full attack. My mother labeled that constellation of events as anger, and from that point forward I could detect the instant my anger circuitry was triggered.

For the life of me, however, I could not understand why anyone would choose to engage with their anger circuitry and let it run, since it felt so violent and unhealthy inside of my body. By paying attention to the early signs of the trigger, I realized that I had the power to control it, by shutting it all down before it blew. However, as time progressed and my Character 2 recovered, I realized that if the circuit was triggered, it took less than 90 seconds for that specific neurological loop to run and then completely dissipate.

Character 2 in the World

As I present various characteristics of our left-brain emotional Character 2, if you are familiar with Carl Jung's archetype the Shadow, you will probably recognize a lot of similarities between the two. The Shadow is often described as the unknown, dark side of our personality, while our Character 2 often presents itself as the unsavory or most deeply pained part of our unconscious left brain. At its worst, this part of our brain is emotionally reactive to the external world and does not accept responsibility for its behavior.

It is also inclined to sacrifice its future, as it is blinded by the pain from its past.

If you are versed in John Bowlby's Attachment Theory, which analyzes a child's anxiety and distress response when it is separated from its primary caregiver, you will also note that many of those negative emotions stem from our left-brain emotional Character 2. All of us, of course, are uniquely wired with both positive and negative emotional circuitry, and how often we run each of these circuits will be impacted by our nature and nurture.

When Character 2 disappeared from my brain, I felt an overwhelming sense of relief and peacefulness. My little Character 2 had embodied a lifetime of emotional pain, and the killing off of her by the stroke was a phenomenal blessing. But of course, I did not end up scot-free, as eventually I regained the ability to emote again, and my new Character 2 has proven to be just as insolent as before. In the big picture, however, it is a relief for me to experience my emotions again, as they not only add depth to life, but they stretch my boundaries for personal growth.

By anyone's measure, I had every reason in the world to be unhappy or feel despair about my post-stroke circumstances, but my right brain felt nothing but gratitude because I was still alive. Although I had fallen completely off the Harvard ladder, which had been the pinnacle of my life's professional effort, I did not feel embarrassment or shame. In addition, with only my right-brain consciousness in operation, I could not understand the concepts of self-loathing, feeling guilty, or feeling lonely. I was not at all depressed, because I did not die that day, which meant I had a second chance at life.

I call my left-brain emotional Character 2 "Abby." I am guessing that my original childhood wound stemmed from a feeling of being abandoned, simply because I was instantly separated from my mother's womb at birth. No matter how romantically anyone paints that picture, physiologically I came flying out of a warm liquid environment where my senses had been muted from sound, light, and touch. Passage from that fluid world, where I felt like a part of my mother's lovely heartbeat, into this cold world of probing, prodding, and sensory overload spontaneously generated a shift in my being that made my whole soul wail. Welcome to the world, baby Abby!

Because our left-brain emotional Character 2 runs all of our experiences through the emotional filter of what has given us pain and what is dangerous, bad, or wrong so it can protect us, it pessimistically sees the glass as half empty. In addition, for the scarcity-minded part of our left-brain Character 2, there is never enough money, love, stuff, or food for everyone. That is why this part of our brain is intent on making sure it gets its fair share, and somewhere within this constricted thinking, our Character 2 can feel discontent and insatiable as it unremittingly yearns for more. Character 2 can feel happiness, but such happiness is based on external conditions and is as transient as all of the other emotions.

It is not unusual for our Character 2 to become bitter or hold grudges or resentments toward anyone who has failed us, and we may protect ourselves from further hurt by being secretive or guarded. We are also good at plotting revenge or slandering. As a result of a self-induced isolation, our Character 2 can get caught up in the dread and doom of a threatening world. When this happens, we need to remember that this is a group of cells that sacrificed our connection with the cosmic flow, so we can sympathize with its inherent value and help it heal before it feels the need to either explode its fears into the external world or cause us to implode with illness.

I know my little Abby is full-force present when I am feeling unappreciated, undervalued, unwanted, or unworthy. I also know that I am stuck in my Character 2 whenever I am feeling oppressed, victimized, or remotely envious. If Abby is feeling stressed out and ready to blow like a pressure cooker, it's never pretty, as I may become impatient or argumentative. And predictably, when Abby is not happy, she does not want you to be happy either.

You know you are dealing with someone's Character 2 if they are being a hateful bully, seeking revenge, being belligerent, using sarcastic humor, or purposely trying to provoke you. This part of our character can be selfish, self-absorbed, self-righteous, self-promoting, or even emotionally manipulative. We may describe a Character 2 as narcissistic, grandiose, pompous, arrogant, self-righteous, or egocentric. On a bad day, this part of our brain enthusiastically belittles others, goes on the attack with abusive name-calling, and spars with a tit-for-tat mentality. It can be critical, intolerant of racial or religious differences, spiteful, mean, or even cruel. And Lord help us all, but this part of our brain is not able to take any responsibility for anything. With a superficial charm, it will love you conditionally, but only for as long as you are letting it control you and you are feeding its needs. Because this part of our character perceives itself as superior, it often has no respect for authority and may behave as though it is above the law.

And when it comes to honesty, our Character 2 is not our highest self. It is tricky and savvy in the way it deceives, and it will absolutely lie to your face and rob you blind. As you might imagine, when it competes it will play dirty and cheat. Our Character 2 has mastered the fine art of deflecting blame, and is seen by others as immature, unsophisticated, dishonest, and unrefined.

Yet in spite of all of these negative traits, we must remember that at the core of this behavior are pain and fear. As I mentioned before, none of us came into this world with a manual about how to get it all right, and if we are to heal this part of ourselves, we must recognize when it shows up, love it in spite of itself, and use our other characters to listen to its needs and assure it that it is both valuable and safe. This is the power of the Four Characters and the goal of the Brain Huddle. When held as an esteemed and valuable member of our whole brain, our Character 2 can feel supported and its reactivity can be methodically de-escalated by the huddle. In addition, just knowing that all of these passionate emotions are cells running in circuit allows us the strategy to purposely and consciously choose to step away from the pain and dissipate that energy.

CHARACTER 2 AT WORK AND AT PLAY

Just as we explored our Character 1 in a couple of different real-life scenarios, let's take those same adventures with our Character 2.

Character 2 on the Job

Regardless of whether Character 2 is a worker or a leader, they are going to have some predictable tendencies in common. First and foremost, Character 2, by nature, does not trust the motivations of others, so Character 2 leaders will not trust the motivations of their workforce and Character 2 workers will not trust the motivations of their leader. As a consequence, Character 2 will lead with an iron fist, make hard demands, and use threats to force production deadlines.

Although this sounds somewhat similar to the leadership style of the Hard Character 1, who leads as though the sky will fall if we fail, our Character 2 leads as though the sky IS falling: so do this now or we will all die, and if you don't die, I'll hurt you when it's over. The Hard Character 1 leader has the ability to slip into the Character 2 bully mentality when it is feeling pressured or pushed up against the wall by outside forces. Anything that it cannot control, whether it is the competition or an external audit, might send Character 1 into Character 2's "throw a fit" style of leadership.

When under stress, a Soft Character 1 leader can devolve into a Hard Character 1 leader, and under the worst of circumstances this leader can then become a Character 2 leader. When you see a Character 2 in the corporate world, they often look a bit agitated or frumpy, as though their clothes are wearing them.

As a worker, our Character 2 may abuse its power by being overly rigid, even when it is clear that a simple solution could be found with a reasonable bending of the rules. Character 2s can make a mess of things simply because they refuse to make rational decisions, even when the situation clearly demands it or failure will ensue. These are the workers who won't stretch a rule when it is clear that now is the exceptional time when the rule should be broken. Also, as a worker our Character 2 tends to take things personally, so any constructive criticism may feel like abuse and deflate any goodwill the Character 2 might have for others.

A Character 2 leader tends to make shortsighted decisions that, if not corrected early, may derail a long-term project. A Character 2 worker cannot handle more than one thing at a time without shutting down or feeling overloaded and overwhelmed. Although our

Character 2 tends to be a perfectionist, it will skip steps in order to fit the timeline or get something done.

Our Character 2 can be dishonest with herself about reality and the consequences of that shortsighted thinking, and facing the consequences of skipping those earlier steps in the project can paralyze her in fear. A Character 2 may skirt the truth with her boss, just as she skirts her own reality, in an effort to minimize her vulnerability.

Character 2 at the Beach

The Character 2 at the beach is worried about the sand because it is everywhere. It's in my towel, between my toes, in my bathing suit, and even in my hair. Also, what's in the water that I cannot see? Is it something that can hurt me, bite me, sting me? It's spooky and salty and buggy here, and I'm restless because I don't feel safe.

The imagination of the Character 2 goes to the worst of the worst beach stories that it has ever heard and brings those stories to the present moment as a real possibility. *Jaws*, anyone? Our Character 2 smells the rotting seaweed and can't help but notice the ugly litter those inconsiderate people left behind. There aren't enough shells, or nice enough shells, and those broken shells are sharp and they hurt my feet. I am niggling with my stuff and keeping an eye on my watch as I pick sand out of my ear.

Our Character 2 cannot relax amid the unknown, so it tries to minimize the number of variables that it cannot control. In preparation for coming to the beach, Character 2s worry that it might rain or that they will burn up in the sun. Will I be able to get the right spot? The neighbor's music is too loud, and the whiff of that cigar is nauseating! I'm worried that I don't look good in my suit, and there's too much wind, or not enough breeze to cool me down from the scorching sun. Those kids over there keep screaming and I can't even hear the waves. My sweat is stinging my eyes, that dead fish stinks and is attracting flies, and the water is foaming. That is so gross! This really is a dreadfully boring place, and I'm probably not even going to see a dolphin today, so when do we get to go home?

Character 2s try to not draw attention to themselves and are critical of how they look, so they wear dull colors along with a cover-up.

They might even appear a bit disheveled since they are uptight and cannot get comfortable, and in their restlessness they are constantly futzing with their stuff. Character 2s prefer to observe rather than participate in activities, so I'm guessing there are no Character 2s playing volleyball or movin' and groovin' on the dance floor. They do like to observe others and enjoy poking fun or being critical of others, and since misery loves miserable company, a Character 2 may have invited another Character 2 along for the trip.

Amid all of this discontent, a Character 2 is rarely self-aware. Preoccupied with their pain or fear, they find it difficult to conceptualize beyond any of the obvious black-and-white solutions to their stressors. When in full force, our Character 2 circuitry is so powerful that it can overwhelm the circuitry of our other characters, leaving them feeling isolated and alone. Unaware of any aid their other characters might bring to the table, our Character 2 will reach out rather than in, becoming dependent on another person for salvation.

Our Character 2 will worry, complain, judge, and perhaps even self-deprecate, all the while being completely unconscious that they are presenting their edginess to others. This can be problematic of course, because although they may attract other Character 2s, the rest of us tend to shy away from this affect and stop inviting them to come along. Sadly, this rejection only reinforces the negative frame of mind of the Character 2, giving them even more evidence to support their negative mindset.

A Snapshot of Your Character 2

- Anger/Name-Calling: *If I get upset and can't resist calling you nasty names, my Character 2 has possessed my brain and is out of control. In these moments it is wise for me to push the pause button so I can give myself 90 seconds for a successful Brain Huddle time-out.*

- Deceives: *When my left-brain Character 2 decides it is going to tell a lie, it tells my right brain to not show our deception in the tone of my voice or in the expression on my face. At this point, my right brain will either cooperate or give me away.*

- Feels Guilty: *If I feel bad for not sending that sympathy card, or for not helping that little old lady cross the street, my Character 2 is online.*

- Internalizes Shame: *When I feel as though I am not good enough, or I am not worthy of being loved, my Character 2 is dominant. It is so important to remember that the Brain Huddle is just a thought away.*

- Loves Conditionally: *If I am generous with my love only when people are doing what I want them to do, this is how my Character 2 loves conditionally.*

- Negative Self-Judgment: *That voice inside that tells us we are not worthy of total joy and everything good that life has to offer is our Character 2. This voice can be really critical and say degrading things to us about ourselves, and when it speaks, it is usually the loudest and trumps all the others.*

- Anxious: *When I am filled with fret and worry about something that could possibly happen, I feel horribly uncomfortable inside my skin.*

- Whines: *Oh no, please, little Character 2, please stop. Brain Huddle, or maybe a cuddle?*

- Egocentric: *To my Character 2, we are the center of the universe, and our needs are all that matter because we are the most important. (Didn't you get the memo?)*

- Blames: *It's all your fault that I am unhappy, or broke, or unemployed, or . . . or . . . or . . .*

Getting to Know Your Character 2

Just as with Character 1, let's explore these questions about your Left Emotional Character 2.

And just as with Character 1, if you prefer to skip these questions for now and continue reading, please feel free to do so.

1. Do you recognize your Character 2? Pause for a moment and imagine yourself engaging in Character 2 behaviors. Picture yourself feeling resentment or jealousy, or whatever your core issues are. A lot of different emotions are masked as anger. Do you have strategies to calm this circuitry, or does your Character 2 tend to seep out into your life in unloving ways?

I know my little Abby very well. She does not show up with hostility very often because I have a lot of roadblocks set up by my other characters to protect me, but if you are determined to poke the bear, she may eventually come out and bite you. Anger is an energetic response to pain, and pain can have a lot of underlying sources. Abby is my hurt from the past, so there are times when I will choose to feel deep grief for those I love who are no longer right here in my huggable space, or she may run those old storylines and fight habitual wars. Abby is not complicated, but rather complex and predictable. She really is a lovely and vulnerable part of me. Do you know these vulnerabilities too?

2. What does Character 2 feel like inside of your body? Do you feel anger, anxiety, or panic very often? How do you hold your body or change your voice when Character 2 comes online? What does that upset feel like inside of your body?

I feel Abby as soon as she takes command of my consciousness. When she is protecting me, she tends to get loud and resonate with a forceful vibration in her voice, and if she feels threatened, she storms my body with a flush of agitation. I feel Abby in the constriction of my chest resulting in shallow breaths. I become alert and move quickly, like a skulking animal on the prowl, and my whole being radiates a palpable discomfort.

This wounded part of me is clearly an innocent child who is using antisocial tools to protect itself (and the rest of me). I have learned that as soon as I recognize Abby has leaped into my consciousness, I need to immediately assemble all four of my characters into a Brain Huddle and address her needs. This tool works for me, as it makes Abby feel safe, heard, valued, and comforted so she can calm down.

3. We have already noted that our Character 2 reflects the Jungian archetype of the Shadow, which is by definition the most primitive part of our brain. Our Character 2 is a part of our unconscious brain that may be unknown or outright rejected by our conscious Character 1. If you tend to keep your emotions under wraps, you may not recognize your Character 2 at all.

Generally, most of us do not have any problem recognizing this part of ourselves or recognizing it in others. But if you cannot relate to your Character 2, I encourage you to speak with those around you to see if they can offer you any insight. For a few of us, this character is our primary personality, as we tend to worry a lot, complain often, and feel as though the world is a treacherous place. If it turns out that this is your primary character, and you would like to experience more joy, getting to know and making room for your other characters will probably help. Hence, training your Four Characters to do the Brain Huddle will help all of them feel like valued members of the team.

If you cannot identify your Character 2, do you ever feel like a victim of circumstances, or helpless to get your needs met? Consider who brings out the best and worst in you. Is there someone in your life you verbally spar with, or do you know a bully who purposely tries to make you feel bad? Is there someone you tend to poke at when they don't feel well, or is there someone who consistently provokes you? Do you get upset about politics or not feel safe around people from other parts of the world? Do you tend to worry?

Our Character 2 by nature is biased against people who are different from us. We feel safe around those who think and feel, and judge what is right and wrong, in the same way we do. Our Character 2 feels safe when teaming up with others who cheer for the same ball team, donate to the same nonprofits, or vote for the same leaders. To our Character 2, what feels familiar feels safe.

Our Character 2 is also the richness of our deepest and most beautiful pain. It is our yearning to be loved, and the depth of our grief and sorrow. Our Character 2 runs the gamut of positive and negative emotions. We are so fortunate to have this capacity so our lives can be enriched and nuanced.

The more we become aware of our Character 2 and its antics for how it expresses itself in the world, the easier it becomes for our other characters to successfully manage its needs. I'm guessing that most of us would like to figure out how we can listen to and satisfy our Character 2 long before it sabotages our relationships or disrupts our deep inner joy. Character 2 tends to explode onto the stage of our lives instantly and powerfully, and with it comes a negative feeling in our gut, a furrow in our brow, a stiff body posture, or an aggressive tone in our voice. Our Character 2 might be bold and loud and mean and biting, or it could be self-loathing, silent, pitiful, passive-aggressive, awkward, or anything in between.

Regardless of whom you might uncover as this character in your left emotional brain, this part of you represents the leading edge of your personal growth. We must master our relationship with our Character 2 if we want to live peacefully within ourselves and with others. Calling on our Four Characters to execute the Brain Huddle is the best way I have found to hold my Character 2 in love and to retain my deep inner peace.

4. Assuming you can identify your left-brain emotional Character 2, do you value this character, or does this part of your brain scare you? How much of the time do you let this character run your life, and under what circumstances?

I have learned to value this character as my inner alarm and for the depth of my emotions as well as my potential for growth. Yet every time this part of my brain goes negatively haywire, it means there is something going on that makes me feel unsafe deep inside. When I am willing to explore what is at the core of my reaction, there are insights available to help me better understand my own fears and weaknesses. Fortunately my other characters know how to self-soothe little Abby, especially if she has come out because she was hungry or tired or because my blood sugar dropped. Nothing helps my Four Characters hold the space for myself better than going into the Brain Huddle.

5. As you think about your Left Emotional Character 2, can you assign it an appropriate name?

As I mentioned, I chose to name my little emotional Character 2 Abby because I believe my original wound came with the inevitable sense of abandonment I felt when I left my mother's womb. We are all individuals, separate from the whole at a physical level, so we are capable of feeling immense loneliness and isolation. No longer completely connected to another, and capable of feeling deep emotional grief and pain, our emotional cells and circuitry make life both enriched and agonizing. It is important for you to pick a name for your Character 2 that is both personal and meaningful to you.

6. Who are some of the Character 2s over the course of your life who have influenced you, in positive or negative ways? Were you emboldened by those encounters or repressed by them?

The strongest and longest relationship I have had with a Character 2 has unquestionably been with my brother, who was eventually diagnosed with the brain disorder schizophrenia. Although we had a tumultuous relationship through our teenage years, as a young adult I consciously chose to purposefully funnel my anger and pain into advocating for the mentally ill. Because of this relationship with my brother's Character 2, as well as the emotional loss of my beloved big brother to this insidious disorder, I was motivated to figure out how I might contribute something positive to help people with mental illness. Although these lessons were hard learned, I would not be who I am today if it were not for my brother's illness and his Character 2.

7. Who in your life appreciates, cares for, identifies with, and wants to hang out with your Character 2? What are those relationships like?

Abby is my childhood pain, and every now and again she likes to get together with other Character 2s to moan and groan till the cows come home. Usually over pizza. No one loved my little Abby better than my mom, as she knew the magical formula that would make me laugh out loud and instantly bring Abby comfort. G. G. and I shared the deep, agonizing pain of losing my brother's brain to his disease, so together we shouldered the burden of his journey in and out of hospitals and jails. We supported each other's Character 2 during our greatest times of need and moments of despair.

Through this experience with G. G., I realized that every time my little Abby called her up for support, she shifted instantly into her Character 4, listened to me, supported me, and nurtured me. Then she would say something ridiculous and totally crack me up, shifting me into my Character 3. Watching G. G.'s success with Abby is how I learned to use my own Character 4, as well as the rest of my characters, to effectively self-soothe my Character 2's pain. Since my mom's death a few years ago, I have perfected going into the Brain Huddle as soon as I feel the need, as I find instant support, camaraderie, and peace there. Unless, of course, I'm in the mood for pizza.

In the big picture of my life, I am very fortunate in that I have a close network of friends who are both loving and forgiving. If Abby does pop out and is unkind or feeling threatened by something, they know how to hold the space for her. Just the other day while I was chatting with a friend over the phone, she just asked me right out if I was in a bad mood. That's all it took for me to realize that Abby was on the line, and I switched immediately into Helen. Learning how to support each other's Character 2 in safe and kind ways is a priceless gift we can give to one another during our moments of need. Learning to gently encourage one another to take a Brain Huddle is a fantastic language for us to share with those we love.

8. Who in your life does not get along with your Character 2?

Any fight between two Character 2s will never be resolved. That statement should be made into a poster and placed in every home and office, and it should go viral on social media. Think about this the next time you pick a fight with someone or they pick a fight with you. If you are feeling disagreeable and ready to go on the attack, pay attention to which character the other person is starting out as. Then notice how your Character 2 uses its power to totally ruin their mood (and perhaps the entire relationship).

In order for any conflict between two Character 2s to be resolved and for any healing or agreement to happen, one of the parties must be willing to shift out of their Character 2. It is really fun to watch this dynamic in action when people are in dispute. Once you gain an eye for it and learn to manage your own Character 2 reactivity, your communication with others will surely become more agreeable.

9. What kind of parent, partner, or friend is your Character 2?

Little Abby is a child, and anyone who parents from the position of their deep emotional pain, their discontent, anger, or immature Character 2 is not promoting a healthy connection. If you are consistently running your Character 2 as a partner, then by definition you are caught up in the circuitry of your pain and misery and loving your partner conditionally. As a result, your partner is probably disconnected or emotionally drained.

The same is true for your friendships. If you are consistently bringing your Character 2 into those conversations, in the form of either your deep pain or hostility, you might want to examine the deeper dynamics of those relationships. No one hurts more or picks a fight, holds a grudge, blames, needs, or criticizes others better than our Character 2. If you are not feeling valued in your relationships, or you feel like your needs are not being met, you might consider bringing all four of your characters into a Brain Huddle for a serious reflection and contemplation about who and how your other characters might jump in to self-soothe you.

10. Again, I don't mean to jump too far ahead of myself, but it's important to consider: What is the relationship like inside your head between your Character 2 and your other characters? Does your Character 2 respect and value your other characters, or does it relish disagreeing with and antagonizing them?

I have spent years working with my Four Characters to help them individually establish healthy relationships with Abby, based on their unique skills. As a result, my Character 2 knows she can count on these relationships during her time of need because my Four Characters effectively practice the Brain Huddle. Hopefully long before my Character 2 really needs it.

The routine you use may look a little different from mine, but if you are willing to go into the Brain Huddle, we will end up with the same peace of mind. For me, when Abby is distraught, my Character 1 jumps right in to make sure she is physically safe, and if there is an immediate problem, Helen takes care of it. At the same time that Helen is fixing the problem, my Character 4 leans in to wrap Abby up in love, because all of my characters recognize that Abby is just a scared little girl who is in pain.

My Character 4 supports Abby by listening to her compassionately. I let Abby know that she is both valued and loved. Perhaps most importantly, however, my Character 4 makes it clear that Abby is not alone and the rest of my characters have her back, especially during her darkest moments. Once Abby has calmed down a bit because she knows she is being supported, held, and heard by my Character 4, and she realizes that the problem is being managed by my Character 1, my little Character 3 can come online and invite her to come out and do something fun. Character 3s are very active, creative, and resourceful, so getting our Character 2 out of their pain and into their body is often a good idea. Before you know it, even the worst of traumas can be handled by our Four Characters, and the Brain Huddle sets us up to live our best life.

Cheat Sheet: Get to Know Your Character 2

1. Do you recognize your Character 2? Pause for a moment and imagine yourself engaging in Character 2 behaviors. Picture yourself feeling resentment or jealousy, or whatever your core issues are. A lot of different emotions are masked as anger. Do you have strategies to calm this circuitry, or does your Character 2 tend to seep out into your life in unloving ways?

_____

_____

2. What does Character 2 feel like inside of your body? Do you feel anger, anxiety, or panic very often? How do you hold your body or change your voice when Character 2 comes online? What does that upset feel like inside of your body?

_____

_____

3. We have already noted that our Character 2 reflects the Jungian archetype of the Shadow, which is by definition the most primitive part of our brain. Our Character 2 is a part of our unconscious brain that may be unknown or outright rejected by our conscious Character 1. If you tend to keep your emotions under wraps, you may not recognize your Character 2 at all.

_____

_____

4. Assuming you can identify your left-brain emotional Character 2, do you value this character, or does this part of your brain scare you? How much of the time do you let this character run your life, and under what circumstances?

_____

_____

5. As you think about your Left Emotional Character 2, can you assign it an appropriate name?

_____

_____

6. Who are some of the Character 2s over the course of your life who have influenced you, in positive or negative ways? Were you emboldened by those encounters or repressed by them?

_____

_____

7. Who in your life appreciates, cares for, identifies with, and wants to hang out with your Character 2? What are those relationships like?

_____

_____

8. Who in your life does not get along with your Character 2?

_____

_____

9. What kind of parent, partner, or friend is your Character 2?

_____

_____

10. Again, I don't mean to jump too far ahead of myself, but it is important to consider: What is the relationship like inside your head between your Character 2 and your other characters? Does your Character 2 respect and value your other characters, or does it relish disagreeing with and antagonizing them?

_____

_____

CHAPTER 6

CHARACTER 3— RIGHT BRAIN EMOTIONAL

In the previous chapter we learned that at the most fundamental level, our emotional left-brain Character 2 determines our current level of safety by bringing in information about the present moment and then comparing that stimulation to threats from our past. Our right-brain Character 3, on the other hand, assesses the threat in the present moment based completely on the information it is processing in the here and now. As a result, our right-brain Character 3 brings a crucial and unique skill to how we process threats. Because our right-brain Character 3 perceives everything as interconnected and in the flow of all that is, it provides us with the big-picture bird's eye view of danger, whether that threat relates to the people around us or to our environment.

When it comes to assessing how safe we are in the presence of another person, our right-brain Character 3 is a well-honed truth detector. It reads body language, matches it with facial expression, and then interprets the emotional cues of inflection of voice and vocal intonation. When all those pieces of the puzzle fit together appropriately, we interpret that behavior as truth. When the pieces do not fit together as they should—for example, if a person's body stance does not communicate openness when they are professing their love—we question the overall integrity of what they are saying.

Some people, for whatever reason, have mastered the fine art of deception, and they do this by consciously manipulating how they are perceived. They fly under other people's right-brain radar. It is possible for us to train ourselves to do this, but to be a really good liar, our left brain has to recruit our right brain to help pull off the trickery. Our right brain would be responsible for holding our body just right and not giving the deception away with our mouth or eyes while manifesting appropriate vocal cues. If for some reason our right brain chooses to not be in cahoots with our left brain's desire to deceive, expect to be caught and pay the consequences.

Our right-brain Character 3 interprets our overall level of safety in our immediate environment based upon how much of what we are experiencing feels familiar. Our right brain is constantly evaluating the bigger picture of where we are, and it is ever assessing—albeit perhaps silently in the background of our awareness—an escape route in the event that we should be cornered. However, it is not unusual for our left-brain Character 1 to step in and override these right-brain self-preserving sensitivities. Although our right brain might alert us to a danger, if we choose to listen to the louder rationalizing voice of our left-brain Character 1 instead, we may unwittingly step into trouble. A fabulous book on this subject is *The Gift of Fear* by Gavin de Becker.

BOUNDLESS AND PRESENT

From a perceptual perspective, when the stroke wiped out the cells of my left brain, my whole world went topsy-turvy. Without those left-brain skills opposing my right-brain experience, I could no longer define the boundaries of where I began and where I ended. Because there was no separation between the atoms and molecules that made up me and those making up everything around me, I no longer perceived myself to be an individual. As a result of this blending, I perceived myself to be a fluid rather than a solid, existing in a constant state of motion and change. Without boundaries, I was exactly that: boundless. So I perceived myself to be both free flowing and as big as the universe.

Think about what that might mean for you. Although you will always have the taming force of your left-brain Characters 1 and 2, this right-brain part of you is always there, too, and always turned on. It is a fine art for us to devolve our focus away from the past or future to attend to the sensibilities of the present moment. When we do this, the details of our life fade away and our experience of the present moment expands.

When my left brain went off-line, I lost all words and language, including the mental file that held all the details of my life. Consequently, I had no identity and knew nothing about myself. Although the consciousness of me still existed inside this same body, the person I had been before, her likes and dislikes, no longer existed. Yet even in the absence of my left-brain ego-self, I remained a conscious and living being. I simply could no longer communicate with words. They had become mere sounds that had no meaning. This experience was the first step in my Hero's Journey, when I laid down the sword of my ego—my individuality—and stepped into the realm of my unconscious right brain.

Have you ever been so scared, excited, or dumbfounded that you could not speak, or time seemed to slow down? Have you ever awakened in a strange place and momentarily forgotten where you were? During such moments we are fully conscious and aware, but there is a brief disconnect from our left brain, with all of its background and reality-based information. Sometimes we are thrust into the right here, right now not by our choosing. At other times we arrive in the present moment by choice.

All we have to do to bring our mind to the present moment is to push the pause button on whatever it is that we are doing, thinking, or feeling and consciously bring our attention to our immediate sensory experience of textures, sights, and smells. This is easy when we are willing to step away from the details of our lives and shift our focus to what life *feels* like. Not how we feel emotionally but how we feel *experientially.* You know how it feels when the sun is kissing your face with its warmth, or you tune in to the vibration of the jet passing overhead. That is the kind of feeling I tune in to when I am in my Character 3. It is not so much my emotions I am in touch with, as those are more the realm of my left-brain Character 2. My

Character 3 is sensually experiential whether I am swimming in the thickness of water or swinging at a tennis ball with an expanded focus on all that is.

> **Sometimes we are thrust into the right here, right now not by our choosing. At other times we arrive in the present moment by choice.**

I know I am in my right brain when I am feeling a sense of gratitude for anything at all, be it my life, my circumstances, or the friendship of another. Joy, however, is the underlying feeling of our right-brain Character 3, so if you want to shift quickly into your Character 3, get experiential: do something fun and engage your sense of humor. And the messier it is, the better! Anytime we laugh out loud, we can't help but be open, present, and completely vulnerable, which is why it feels so great and is so good for us.

After my stroke, because I had lost all perception of time, I existed in the eternal timelessness of the present moment. To my mind, the linearity of time was no longer measured in the man-made intervals of seconds, minutes, and hours. Instead, time passed in moments: some brief, some longer, all depending on what I was doing. Playing or creating in the absence of my left brain's judgment was both meaningful and satisfying.

Without any awareness of my physical boundaries, it was impossible for me to distinguish other people as entities that were separate from me. As a result, I perceived us all collectively and energetically as a part of the same being. It was as though we were all woven together into a fabric of tiny molecules that were in motion, and together we made up a tapestry of humanity. We didn't need to communicate with words because we empathically felt one another's feelings, and we communicated through facial expressions and body language. We were all in the flow together, as a single unit that was made up of the sum of our parts.

This is comparable to what happens when we get caught up in the right here, right now excitement of a ball game. We shift into our Character 3s as a group while we sit together on the edge of our seats, being blown away by those incredible plays or those jaw-dropping volleys. Our consciousness collectively expands, and as a whole we leap to our feet, high-five, scream with elation, and maybe even do the "wave" as a fused group. Caught up in the moment, this is not about me and it is not about you. It is about all of us as one team, and our energy has the potency to bust the roof off the stadium. How fantastic it is when we are here together, sharing these amazing moments. Lost in the flow of the excitement, we are each a part of the whole, all dressed in the same colors. *Gosh, that game flew right by. I can't believe how late it is. I am so hungry.* This is time well spent in our right-brain Character 3s.

Sometimes I think about how a culture of bacteria that is made up of a bunch of individual cells has the ability to collectively share enough of a communal awareness that they can work together to infect and overtake a host. Even though each of those cells is an individual, they work in synchrony to become a powerful predator against a body that is billions of times their size. This is comparable to the trillions of cells that make up your body, with each cell an individual with its own position, shape, and job. And somehow, all of those cells work independently to do their part and then collectively they communicate with one another, making up a healthy you.

This is how we humans, as a species, exist and function when we are in the consciousness of our right brains. We are all equally important brothers and sisters who are united as one human family. Our uniqueness contributes to the betterment and increased variability and viability of the whole. I believe that our right-brain emotional Character 3 is the same part of ourselves as Carl Jung's archetype the Anima/Animus, which represents the inner femininity of males and the inner masculinity of females. According to Jung, all humans are energetically androgynous, and this part of ourselves serves as the primary source of communication with the collective consciousness of our species, independent of our gender. When we are all joining in a high five, it does not matter what gender we are.

When it comes to the magnificence of humanity, our right brains are clear that our differences contribute to our creativity and versatility. But unfortunately our left brain's tendency toward negative judgment toward others who are different from us sets us up for separatism, racism, and bigotry. The truth is that our strength is in our differences, not in our similarities. If you were stranded on a desert island, would you want another person just like you around or someone who had different interests and a different set of skills? If I were stranded on a desert island, I would welcome your differences, and my left brain would immediately drop its tendency toward superiority and negative judgment, or else you might just toss me off the island or make me fend for myself.

Emotion, Past and Present

As we discussed at length in the previous chapter, the fundamental anatomical difference between our left-brain emotional Character 2 and our right-brain emotional Character 3 is how these two emotional groups of cells process the information they receive from those reptilian brain stem cells. Briefly, our left-brain Character 2 amygdala will immediately compare incoming information about the present moment with memories from our past. As soon as this happens, the consciousness of our left-brain emotional Character 2 shifts out of the present moment and linearly processes incoming stimulation about the external world. As a result, if I am to feel remorse or guilt, or hold a grudge, I feel these emotions in the present moment but they are about something that happened in the past. It is the circuitry of my left-brain emotional Character 2 that has the ability to experience dozens of specific emotions, both positive and negative, but they are relative to past and future experiences.

Simultaneously, our Character 3 is the consciousness that has evolved in our right-brain emotional tissue, which experiences emotions in the right here, right now about this present moment. Because our Character 3 has no perception about the past, and it never disconnects from the consciousness of the present moment, it always exists at the level of the flow of the universe. Call this the consciousness of the One, God, Allah, the power of the present moment, Nature, the Universe, whatever suits your belief system. Our right-brain consciousness is the realm of that unconscious dimension that is ever flowing in the background of our left-brain focus on the external world.

What this means is that when we feel that we are alone, it is because our left brain perceives, feels, and experiences us to be alone. But when we release our attachments to people and things in our external reality, we shift back into the consciousness of the flow, whereby we can experience gratitude and joy. At any moment we have the ability to choose which consciousness we want to focus on: our left-brain external reality or our right-brain present moment. It is one or the other at any moment in time. We are either focused on our individuality or we allow ourselves to blend into the flow.

In many instances when I am in my right-brain emotional Character 3, it is difficult if not impossible for me to use words to describe exactly what I am feeling, because it is impossible, by definition, to use words to describe that which is not describable. For example, when we view art or hear music, our right brain may be moved as we feel that something is beautiful. Our whole soul might be wrapped up by the awe of our existence during a sunset. Or when we stand at the top of a mountain, we can simultaneously feel that we are both as big as the universe and as irrelevant as a speck of dust. These are the moments in our collective consciousness that we can neither measure nor define, yet we share an underlying knowing of what they *feel* like deep inside. It is our Character 3 that feels that magical connection when we feel *home* in a hug.

Creatively, if you are a natural musician or visual artist, you use your right brain to express yourself. When our Character 3 comes out and is dominant, we become uninhibited by the paralyzing fear of our left-brain judgment. It is right here in the present moment when we find a beat, add a rhythm, pick out a melody, and hook up with our left-brain lyric-writing self to communicate a message that is a perfect combination of story, emotion, and feeling. But when we work it, study it, and perfect it, our left brain is the master of practice, and then our right brain makes magic happen during the performance.

**When our Character 3 comes out and is
dominant, we become uninhibited by the
paralyzing fear of our left-brain judgment.**

Many of us compulsively express ourselves artistically, and few things are more beautiful than the feeling we get when we are completely lost in the flow. Some people proclaim that their creative process is an agonizingly painful experience that is, in its own weird way, uniquely delicious. Others just link up to their muse and genius pours right through. I know that when I carve stone, I become so caught up in the flow that I feel compelled to discover and release whatever figure is stuck inside the block. How lovely it is that we humans have this capacity through our right brains to reach deep inside and express ourselves in creative ways. What a bonus it is if our creation somehow touches the heart (or right brain) of another.

This alternate reality of our right brain whereby everything is related is an actual consciousness. But because we cannot define it, see it, touch it, smell it, taste it, or hear it, this parallel world of perception is often minimized, invalidated, and denied by our left-brain counterpart, which only believes in the external world. It is in this realm of the right-brain energetic flow where synchronicities commonly unfold. Yet in the *real* world, for a strong left brain, those synchronicities are easily dismissed as mere coincidences.

This is a fair judgment from our left brain if we consider how threatening this idea of connection might be to the individuality of our left-brain ego-center. The only problem with denying the duality of our left and right brains, and the realms they independently navigate, is the billions of things in our right-brain world that defy our left brain's definition of truth. Even the existence of life itself is unexplainable by our left brain. It is so important to realize that simply because our left brain has an opinion about something, that does not make it true.

When infants are born, their brains have not yet had a chance to define the boundaries of where they begin and end. Consequently, the consciousness of our right brain is dominant at birth and until we have acquired enough information about ourselves and the world around us to establish ourselves as individuals who are separate from the flow. Kids tend to exude their right-brain Character 3s onto the playground with full force until their left brain develops a *reality-based* consciousness, at which point they begin to mature both scholastically and physically. Schools promote our left-brain development, especially with the introduction of reading, writing, and math. Add to that classes like geography and history, which require our left brain to mature so we can memorize a bazillion details. My little right-brain Character 3 never understood why I had to stuff all those dates and details inside my head. Wasn't it enough that I knew where to find them?

When I was around eight years old, I asked my mom if she thought in words or pictures. She told me she thought in language. Thinking in words was a profound concept for me because my brain flashed videos, not letters. Later as adults, G. G. and I vacationed together and shared trashy novels. I would ask her what the book was about and she would give me a rudimentary plot. It turned out that her brain actually read words, just words, while my brain read words and then created a video of the story. One of my favorite books on this subject is *Thinking in Pictures* by Temple Grandin.

When kids play together, they hang with others who have shared interests: "Kickball, anyone?" Our Character 3 will swing on that swing higher and higher, as high as it can get, and it is in those wondrous moments of soaring with the birds that no one is thinking about that spelling test tomorrow. Whatever our age, our Character 3 is the part of our brain that loves doing stuff in our body and being active. It is also the big-kid part of our brain that never grows up, loves to walk in the rain, and catches the ESPN sports recap of the day for all of those spectacular plays.

CHARACTER 3 IN THE WORLD

Our right-brain emotional Character 3 is like a puppy that is always watching every move you make, ready to pounce the moment you reach for the leash, toy, or food bowl. Our Character 3 is the virtuoso who is eager to practice for hours and hours on end at whatever brings it joy and thrills its soul. This part of our brain sees possibilities instead of limitations. Everything is about the connection with myself or with others, and as I practice, practice, practice, I am constantly tweaking and correcting until it *feels* right. What can I do in this moment to strengthen my stride or deepen my breath so I can get a better outcome?

Our Character 3s are witty and hilarious. We will laugh so hard that we stomp the ground as we gasp for more air. Wide open emotionally and in the present moment, we become a pack and feel camaraderie with everyone who is joining in. When we get excited together, we share a deep connection, and these become the moments we cherish, talk about, and later embellish. We thrive as a collective, connected by our similarities and overlooking our differences.

Although our Character 3 is a whole lot of wonderful, it also has the potential to get us into serious trouble. It is Character 3's nature to act impulsively in the present moment without considering the consequences of our behavior. "What were you thinking?" Well, clearly, I was not thinking. I was feeling, as I was having an experience in the present moment and it seemed like a good idea at the time. Unless I am a teenager whose brain has not yet completely developed (and that is a whole different book), suffice it to say that it is the nature of Character 3 to push limits, buck authority, and ask for forgiveness rather than permission, regardless of age.

Our Character 3 is often not interested in conceding to the authority of our Character 1, even though our Character 1 may be very interested in controlling what our Character 3 is up to. I spend a lot of time on a lake, and when the storm clouds roll in, my Character 1 knows very well that it is time for me to head off the water. My Character 3, however, has a mind of her own. She figures those clouds *might* mean a big storm, but they also might pass by me. So until I see lightning, I'll keep hoe-dee-doing this way, and I'll head for cover if the rain begins. I hate to admit it, but only in the last few months have I been willing to let my Character 1 take charge of these sorts of safety decisions. I have been really glad about this several times in just the last few months, and I have to say, *Go Helen.*

Character 3s are quite compulsive and like to do things their own way, and when they have a vision of what they want to accomplish, it is hard to give them any input or suggestions. My Character 3 has adopted the mindset that if you want to get something done quickly, it is easier to just do it yourself than to try to find the right words to tell someone else how you want it done. Character 1s are much better at using their words, as they are methodical and have a pronounced mastery of language. Character 3s, on the other hand, just dive in and go-go-go. Act first and explore the possibilities, and then step back and hope it all went well. The last time I cooked eggs, I was in the mood for some sweet potatoes too. Instead of asking my friend who is a good cook for suggestions, I just dove right in and failed miserably. Sadly, the worst thing you can do when I am in that frame of mind is to try to help me. Mark Twain had it right when he said something about how there are things you can learn by carrying a cat by the tail that you can learn no other way. That would be our Character 3.

CHARACTER 3 AT WORK AND AT PLAY

Character 3 on the Job

Our Character 3 is happy when people are together, so regardless of whether they are a leader or a worker, Character 3s enjoy meeting with people face-to-face so they can *do* things. Character 3s love group projects, but they also work well alone. They pivot easily between different parts of the project and rarely choose to start at the beginning or work linearly. Character 3s thrive on creative projects that are not well defined, and they look for reasons to use space and collaborate.

Character 3s shy away from giving a boss a timeline for when they will be finished with a project because they get all caught up in the doing of the task, and that pesky clock only interrupts the flow of the creative performance. Character 3s will get the job done and probably create something magical, but they prefer to not be held accountable to a deadline. The worst thing any boss can do to a

Character 3 is to ask them to set a plan, a schedule, a deadline, and a budget. And of course, Lord help your board meeting if you ever give a Character 3 a whiteboard and a palette of colorful markers or put them in charge of running the agenda.

Character 3 at the Beach

Character 3s are so excited to get to the beach that they forgot to bring the sunscreen. The towel is tossed in a crumpled wad in the sand, but that is okay because they will dry off in the sun. Dressed in a comfy, colorful Hawaiian shirt and shorts with a cap that doesn't match, they dash along the edge of the water squealing like little pigs because it is a bit nippy and they are so happy to be there. Just look at the way the sun dances through the rippled surface, making a bright, shiny neural network on the ocean floor. It is so beautiful!

Character 3s didn't plan much for this adventure because their anticipation of what fun they would have was so exciting they couldn't focus on anything else. They grabbed whatever they could find to wear, and right here, right now they are busting out their best dance moves and laughing together in total joy. It is such a thrill to see people they already know, and to connect with new friends who have familiar interests and energy. They focus on what they have in common and what they like about one another. Mostly they are grateful they are all here together.

To the Character 3 the mess is an integral part of the beach experience. They relish the sandy beachfront as the perfect play-ground because they are sensual beings and this environment is so rich with the grit, the sun, and the wind. Character 3s are out socializing and saying hello, not just to the people but to the birds and crabs and all those little creatures that scurry. They wear a smile and offer an open invitation for all other Character 3s to join in as they build a sandcastle or bury a pal. In groups, Character 3s play games and make up new ones. They pay the locals to braid cornrows into their hair and then leave them in way longer than they should.

Character 3s celebrate all that the beach has to offer without comparing it to the last time they were there. They pick a sunscreen based on the way it smells or how cool its label is rather than its brand. And rest assured they will probably get sunburned anyway because they will forget to reapply it, if they bother to put it on at all. Oh yes, all that rotting seaweed and those smelly fish over there—what a great place to explore. And those tiny little holes that bubble when the tide recedes, are there creatures in there? Let's dig and see.

The perfect day at the beach for a Character 3 is any day they see a dolphin at play, and they'll hunt for shark's teeth and give them all away. Come sun or come rain, any sensual time in nature without a schedule is an awesome day. Oh my gosh, oh my gosh, that was the BEST time ever. Can we do it again tomorrow?

A Snapshot of Your Character 3

- Forgiving: *We are interested in connecting with others in the present moment and are willing to forgive easily so we can reconnect with them at the heart level.*

- Awe-Inspired: *We exude excitement about everything that is happening in the right here, right now because life is such an amazing gift and every moment is filled with amazing possibilities.*

- Playful: *We are radiating life and every moment is exciting. Being alive is so delicious that we just want to eat up every experience, and there is nothing better than sharing a playful moment with someone else.*

- Empathic: *We are so connected to one another that I can feel your joy as well as your pain. I am able to stand here by your side because your pain does not scare me. I am connected to you. I care about you. I love you, and we are never alone.*

- Creative: *If I take this and I do that with it, I will come up with something completely new. That is so cool. Want to help me?*

- Joyful: *I just want to laugh and laugh, and play, and go get a good adrenaline rush—want to join me?*

- Curious: *Let's explore this, and let's go try that, and did you catch that clue? I wonder where it leads.*

- Style: *I will be wearing my favorite striped top and comfy plaid bottoms, or one of my favorite brain-branded shirts, for sure. Matching? What do you mean, matching?*

- Hopeful: *No matter what, I am here with you and we will get through this together. It will all be okay. I got your back, whatever happens.*

- Experiential: *I love how different experiences feel inside of my body. I am very sensitive to my physiological response to what is going on in my life, and I listen to my gut and intuition.*

Getting to Know Your Character 3

If you prefer to skip these questions for now and continue reading, please feel free to do so. Focusing on these different parts of your brain can be taxing, so approaching them when you are feeling patient and refreshed might be a good idea.

If you're ready now, let's explore your right-brain emotional Character 3.

1. Do you recognize your Character 3? Pause for a moment and imagine yourself being in this present-moment character. Let your left brain drift into the background as you bring your attention to the right here, right now and explore the immediate sounds, textures, sights, and smells. How easy is it for you to make this shift?

Character 3 is my primary character in that I wake up in the morning as this part of myself and then consciously shift myself into a different character as needed. I feel joy in my heart as soon as I wake up, and I am curious about what is on my schedule for the day. Then once I check out my scheduled commitments, I meander from one thing to the next pretty much without a plan but problem solving as I go. At least until Character 1 pops in and I am back on the schedule again. My automatic impulse is to slip back into the freedom of the present moment, unless there is some reason why I should be a different character.

2. What does Character 3 feel like inside of your body? Do you feel your heart expand? Do you stand more on your tippy-toes as though you are lighter? Does your voice disappear because it's not about output but just about bringing it all in? What does your Character 3 feel like when you are having an experience right here, right now?

My Character 3 is a joyful little character who loves life and loves you. It feels very alive inside of me and infiltrates every molecule of my being. My Character 3 is light in my body, healthy, physically strong, and athletically agile. This bright, effusive, uncomplicated, uninhibited, often unruly, explosive part of me expresses herself with reckless abandon.

3. But what if you don't recognize your right-brain emotional Character 3?

If you do not recognize your Character 3, you are missing out on the expression of a whole lot of spontaneous energy that has no plan and no timeline and has limitless curiosity. This expression of ourself is the unbridled emotion of the present moment that might manifest as an uncontrollable belly laugh—or an abrupt explosion of anger.

Right here, right now our Character 3 supports a personality that is exuberant, joyful, and completely devoted to the sensations and experience of this moment, without fear or judgment. It does not know about the past and has no perception of the future. As such, it does not perceive risk as a negative thing but merely as a great adventure and a yummy adrenaline rush. Our Character 3 is emotionally connected to other people through empathy, likes variation, and thrives on anything experiential.

As you might imagine, however, this uncontrolled, unpredictable energy of an unbridled Character 3, which has little if any respect for authority, would probably drive any self-respecting Character 1 bananas. In our society Character 1s are the voice of authority and are often uncomfortable with the impulsive nature of our Character 3s. Consequently, if you do not recognize your Character 3, it may have been driven into submission and quiet obedience when those left-brain characters had little use for its happy-go-lucky, fun-loving, high energy.

4. Assuming you can identify your right-brain emotional Character 3, do you like how this character expresses itself inside of you? How much of the time do you let your Character 3 run your life, and under what circumstances?

I am madly in love with the way my Character 3 feels inside of me, and although I value all Four Characters, I spend most of my time beaming Character 3 into the world. I express this part when I am carving limestone or creating something wondrous out of stained glass. I enthusiastically love to do all kinds of things, and whether it is a dirty job and I am laboring and sweating or I am biking, rowing, swimming, or hanging out with a kindred spirit, my Character 3 feels healthy and ebullient.

Whatever this part of me is up to, you can rest assured that I am creatively scheming in the right here, right now and innovatively participating with every ounce of my being. My Character 3 does not hold back at all, and fortunately my Character 1 keeps an eye out for her well-being when I am lost in the flow. I have learned that a strong Character 3 who is supported by a strong Character 1 is a really productive and beautiful collaboration.

5. As you think about your Right Emotional Character 3, can you assign it an appropriate name? In addition, now that you are more familiar with your left-brain Characters 1 and 2, are you satisfied with the names you assigned to them earlier?

I have named my Character 3 "Pigpen." Remember Charles Schulz's *Peanuts* character who is always walking around in a dust storm? My little right-brain emotional Character 3 is just like that. She is here, she is there, she is everywhere, always in the present moment stirring up some sort of a mess, with gusto. Not so much out of control or embarking on illegalities but definitely out of alignment with the more prudent values of my left brain.

Pigpen is totally open, emotionally available, and ingenious, and in her innocence she is both vulnerable and naive. Because she exists in the right here, right now, if my left brain is not informing me intelligently, my Pigpen has the capacity to make some really bad decisions based purely on ignorance.

I encourage you to reach deep inside for a very special and appropriate name for your Character 3. What fits your joyful and fun-loving nature?

6. Who are some of the Character 3s who have influenced you over the course of your life, in positive or negative ways? Was your Character 3 emboldened by their Character 3 or repressed by it?

My dad, Hal, was a powerful Character 3, so I grew up with my eyes wide open to the advantages and disadvantages of expressing this character. Because Hal was incredibly creative, when I was a child we always had the most phenomenal Halloween outfits you ever saw. And talk about a music man: he could pick up any wind instrument and within 20 minutes be taking song requests. The downside of having Hal as a Character 3, as you might imagine, was the burden this placed on my mother, G. G., to maintain order in the house. As a result, our basement and garage were embarrassing disaster zones and it was impossible to locate anything.

I learned that if I was going to let my Character 3 run wild when in project mode, it was vital that my Character 1 come on board to manage the mess and recreate order regularly. To this day I have a very satisfying relationship between these two characters. If Character 1 didn't come along and do her thing, the total lack of order would render my Character 3 paralyzed and unproductive.

7. Who in your life appreciates, cares for, identifies with, and wants to hang out with your Character 3? What are those relationships like?

I think most of my friends enjoy playing with Pigpen, and they know exactly what they are signing up for when we get together. When I think about my closest friends, we are a creative bunch.

Although Pigpen is extremely patient during projects and versatile in how she likes to spend time, it takes a lot of effort for her to hold the space for someone else's Character 2 until that person is willing to shift into their present-moment joy.

The fine art of matching someone else's emotional needs is a natural skill for Pigpen, as one of Character 3's natural abilities is to feel empathy. Pigpen's aptitude for love and compassion is off the charts, but there are times when my Character 1, Helen, wants to jump right in and fix the problem, or my Character 4 (whom you will meet in the next chapter) steps up to wrap herself around a hurting soul.

Little Abby, my Character 2, also has the capacity to be a good friend to someone else's Character 2 because sometimes life simply boils down to two Character 2s holding space for one another. That famous quote "A friend in need is a friend indeed" speaks loudly to all four of my characters. At the core of who I am, I believe our number one job is to love one another.

8. Who in your life does not get along with your Character 3?

Anyone who is unhappy and committed to staying that way will definitely find Pigpen irritating. Believe it or not, I had a boss who told me once that I could not grow up to be a serious scientist because I was too happy. Sadly, he was suffering emotionally because he was in chronic physical pain. His Character 2 was his daily dominant character, and often when our Character 2 is not happy, we don't want anyone around us to be happy either. A few years later, when I won the Mysell Award in the Harvard Department of Psychiatry, the most prestigious award that department offers to an M.D. or Ph.D. for their research, I reminisced about that negative assessment of my Character 3, and my whole self felt vindicated.

Although my Character 1, Helen, was always in charge of getting me to work on time, my lab-rat life was dominated by the playful spirit of my Character 3. In fact, when I applied for the position in the Harvard Department of Psychiatry, I told my future boss that I was an artist in my heart but I had chosen science to make a living. I essentially had told her that I had a strong Character 3 that was creative, innovative, and exploratory, but I also had a strong Character 1 that would do the job well and meet my deadlines. She hired me immediately and was intuitive enough to assign me the research projects that would benefit from an aesthetic eye. We had a successful working relationship because we capitalized on our individual strengths.

9. What kind of parent, partner, or friend is your Character 3?

Our Character 3s are a fantastic part of who we are, but since they thrive on chaos, the creative process, and the present moment, they may be the most fun and emotionally available parent but are certainly not the most organized or disciplined. It's also important to note that if you are a parent and a primary Character 3, it is neither fair nor appropriate for you to push a child into their Character 1 prematurely so they can bear the burden of creating order.

Children are biologically children and need to be protected as such. A parent who has a drug or alcohol addiction cannot show up as a healthy Character 1, so responsibilities often fall to the oldest child, demanding they prematurely develop their Character 1 at an early age. It is important that we pay attention to the demands we are making on those around us, and this is one of the reasons why this material is so important. Even if you are a primary Character 3, you can train your Character 1 to come online when it is appropriate for you to show up as a healthy adult.

Similarly, it is important that as adults we provide structure not just for our young children but for our teenagers. The human brain does not fully mature until around the age of 25, so although young people may look like adults, until their brains reach maturity our Character 1s need to help them out by providing structure and taking on the role of Character 1 for them. Clearly some young children are born with a penchant for order and perfectionism—in other words, some children start out with some Character 1 skills —but for those who are not born this way, we need to provide structure for them. Although it is important that we befriend our children, it is more important that we parent them.

If you are a Character 3 as a partner, hopefully there is a Character 1 around, or your home may be in total chaos as a result of hoarding. The human brain that has no order may be brilliant, ingenious, creative, innovative, and all those magnificent traits of a Character 3, but without some semblance of order we don't have a neuron to hang our ideas on and thus end up accomplishing very little.

In the meantime, love those little Character 3s, and let them remind you what it was like to be a kid. It will be good for your heart, both physically and emotionally.

10. Although we have not yet fully examined Character 4, it is important to consider how kind the relationship between your characters inside your head is. How does your Character 3 relate to your other characters?

As I have expressed, my Character 3, Pigpen, appreciates my Character 1, Helen, and is completely cooperative with her because she knows that Helen is eager and willing to take care of all the things Pigpen has no interest spending time on. Although Pigpen is really bright, resilient, and creative, she is not an intellectual, thus she finds memorizing details and reading manuals painfully tedious. Thank goodness Helen is here to run our world so Pigpen can get lost in the flow of whatever has most recently caught her fancy. Pigpen has been known to rebel against a schedule, and she does not like to be controlled by others, but when she feels accepted and valued for her natural skills, Pigpen is a loyal, faithful, and committed friend to everyone, especially my other characters.

My Character 3, Pigpen, and Character 2, Abby, also have a really important relationship. When Abby brings a sense of fear or unhappy discontentment into the present moment, Pigpen has mastered the skill of helping Abby find her way out of the pain and into the play. And when Abby is caught in the grip of her deepest inner grief or sadness, Pigpen does not shy away from her pain. Instead Pigpen is a really good friend. She will not only hold the space and comfort Abby but will remind her how blessed they are to be alive, and how important it is that they celebrate this capacity to feel the depth and deliciousness of that misery.

Cheat Sheet: Get to Know Your Character 3

1. Do you recognize your Character 3? Pause for a moment and imagine yourself being in this present-moment character. Let your left brain drift into the background as you bring your attention to the right here, right now and explore the immediate sounds, textures, sights, and smells. How easy is it for you to make this shift?

_____

_____

2. What does Character 3 feel like inside of your body? Do you feel your heart expand? Do you stand more on your tippy-toes as though you are lighter? Does your voice disappear because it's not about output but just about bringing it all in? What does your Character 3 feel like when you are having an experience right here, right now?

_____

_____

3. But what if you don't recognize your right-brain emotional Character 3?

_____

_____

4. Assuming you can identify your right-brain emotional Character 3, do you like how this character expresses itself inside of you? How much of the time do you let your Character 3 run your life, and under what circumstances?

_____

_____

5. As you think about your Right Emotional Character 3, can you assign it an appropriate name? In addition, now that you are more familiar with your left-brain Characters 1 and 2, are you satisfied with the names you assigned to them earlier?

_____

_____

6. Who are some of the Character 3s who have influenced you over the course of your life, in positive or negative ways? Was your Character 3 emboldened by their Character 3 or repressed by it?

_____

_____

7. Who in your life appreciates, cares for, identifies with, and wants to hang out with your Character 3? What are those relationships like?

_____

_____

8. Who in your life does not get along with your Character 3?

_____

_____

9. What kind of parent, partner, or friend is your Character 3?

_____

_____

10. Although we have not yet fully examined Character 4, it is important to consider how kind the relationship between your characters inside your head is. How does your Character 3 relate to your other characters?

_____

_____

CHAPTER 7

Character 4— Right Brain Thinking

Welcome to our Character 4. I say *our* because this is the part of our consciousness, our right thinking brain that we share with one another, and all other life. I see the brain cells underlying our Character 4 as the portal through which the energy of the universe enters into and fuels every cell of our body. This energy and its consciousness fill up our entire being. We are swimming in it, and it is swimming in us. There is no separation. Our Character 4 is the all-knowing intelligence from which we came, and it is how we incarnate the consciousness of the universe.

Although we are an amazing form of life, we are atoms and molecules in motion. As I mentioned in the chapter on Character 1, our left brain refined our level of processing such that we could lift our focus up beyond the level of the molecular flow into the realm of external *things*. However, when we devolve our perception from the level of *things* back down to the level of the atoms that compose all things, we return our focus to the level of the particulate matter from which we originated. The consciousness of this microcosmic flow remains omnipotent and omnipresent. We never left it, we are never without it, and it is the river of peace that flows through our veins.

We embody this peace when we tune in to the consciousness of our Character 4. However, in order to do this, we must quiet the thoughts of our Character 1, which is obsessed with the details of life in the external world. We must settle the emotional volatility and reactivity of our Character 2 and distract our focus away from the experiential sensations being processed by our Character 3. These three characters make a lot of noise inside our head, and we need to quiet it all down if we are going to expand in the consciousness of our Character 4.

I think of the four different consciousnesses of our Four Characters as though they are different instruments playing in a string quartet. The two violins play the melody and the piercing high notes soar above the other sounds, making them easily heard. The cello brings in a bass line of supportive low tones that is readily distinguishable from the higher notes of the violins. The viola, however, plays notes that are not quite as high as the violins or as low as the cello. This is why these middle tones of the viola can blend so well with the others that they become difficult to identify, while at the same time they form the glue that melds all the other notes together into a balanced sound.

> **I think of the four different consciousnesses of our Four Characters as though they are different instruments playing in a string quartet.**

Although the viola may be difficult to hear when everyone is playing, if it were not there, the unity of the overall sound would lack luster. At the level of the Four Characters, the role of the viola represents the consciousness of our Character 4. The two violins symbolize our left-brain Characters 1 and 2, which can be loud and overshadowing, while the cello denotes the lower body tones of our Character 3. We must listen carefully to hear our Character 4, and when all the other instruments agree to play more softly, the strong, exquisite tones of the viola's voice can be heard. Our Character 4, like the viola, is the glue that balances the expression of our Four Characters.

If your Character 1 is having a hard time wrapping its mind around the validity or existence of the consciousness of Character 4, I recognize that it is natural for Character 1 to possibly judge the unfamiliar, unknown, and mystical as *woo-woo*. Yet throughout the history of man, and in varied cultures around the world, humans have designed techniques, tools, and strategies ranging from religious dogma and prayer to meditation and yoga to access this realm of consciousness and experience their Character 4. Carl Jung spoke of the Self as the archetypal part of ourselves that was the unification of our unconsciousness with our consciousness. We know it is there, but how to access it has been the challenge, and each of the characters goes about this in different ways.

Because our left brain thrives on separating and dividing everything into categories so it can create order and make sense in the external world, it decided that science and spirituality are two subjects that are so polarized that they both cannot exist. As a scientist I never understood this frame of mind, because science is the strategic tool that we use to explore what we do not understand, and clearly we do not understand the realm of our right brain.

Unfortunately the scientific method, which good scientists are supposed to use to perform quality research, is by definition a linear-based method that the left brain has designed so it can measure everything and replicate an experiment in order to test a hypothesis. This is quite limiting, of course, since the scientific method as it currently stands is only available to prove and validate things in the external world of our Character 1. We can only use a linear technique to study linear phenomena.

If something is not measurable or if experimental results cannot be replicated, our left brain will often either choose to deny its existence or negate its value altogether. This gap between what we can study in the conscious realm of our left brain, versus everything else that is neither measurable nor replicable in the realm of our right-brain consciousness, requires that we take a leap of faith if we are going to wrap our minds around understanding it. It is reassuring to note that a lot of really creative research is currently going on that is both stretching the boundaries of the scientific method and building a bridge between the dogma of science and the experience of spirituality.

The consciousness of our Character 4 is our constant companion, as it is the energy within which we exist. This consciousness is woven into every cell of our body and every molecule in the universe. It is the energy ball within which we live and breathe and have our being. It is the source of our life, and it is the experience we yearn to achieve through various practices. Our Character 4's consciousness is the most distant destination of our Hero's Journey, and returning to this consciousness is a coming home to our precious and most peaceful self. Our Character 4 is our authenticity, as it is the part of ourselves that we share with the One. Yet that does not negate the understanding that each of our Four Characters is individually authentic.

From Cosmic Energy to Human Life

Our Character 4 is the original consciousness that we were born with, before our brain and body were neurologically wired up for function. Long before our infant brain could define where the boundaries of our body began and ended, we were just a ball of energy infusing and radiating around a lump of cellular life.

Following our conception when our father's DNA combined with our mother's DNA, that single zygote cell would develop into our fetal body. It embarked upon its maturation toward human life, powered by the energetic consciousness of the universe. That individual zygote cell contained the molecular genius necessary for it to metamorphize into who we are today.

Over the course of a nine-month gestation, that ball of cosmic energy (the consciousness of our Character 4) directed the expression of our genes, which contained the blueprint of our molecular profile. As a result, the cells that would make up our form developed at a rate of 250,000 new cells per second. (Yes, per second, not per minute!) As impossible as that may be to imagine, that universal power of our Character 4 consciousness directed our transformation from the single zygote cell into the structure of our emerging body.

By the time we completed a normal nine-month gestation, all of the cells making up our tissues, organs, and organ systems were lined up in perfect formation and ready for our next stage of development that would occur when we left the womb. By the time we were born, although those trillions of cells making up our brain and body were structurally organized and in position, they existed at various levels of function. For example, the cells of our diaphragmatic muscle of respiration were already wired up with our reptilian brain stem cells, so at the time of our birth we were able to breathe. Our skeletal muscles, however, and the rest of our motor system, although set up and in position, would require external stimulation to mature.

At the time of our birth, the ball of energy surrounding the mass of cells that we had developed into while in the womb blended fluidly with the energy of the external world, as they were one and the same. Of course, the journey of our birth from that warm liquid womb into the oxygen-rich gaseous air was by definition not only shocking to our biological system but was also our original separation from our connection with something that had been both protective and nurturing.

In the moment of our birth, we gained our physical individuality, but we would never shed the shared consciousness of the universal energy that infused our every cell. Why do we love babies? Perhaps it has something to do with how easy it is for our Character 4 to be drawn out as we mirror the consciousness of their Character 4. All you have to do is look into the eyes of a new soul and see its beauty. When we unfold their little hands with our fingers and smell the scent of their heads, it transports us into a remembrance of our own innocence, vulnerability, and innate wholeness. With birth we celebrate the miracle of life and this wondrous transformation with an eternal hope for a future of all humanity that is wide open with possibilities.

When we are born, in response to our new environment our infantile brain physiologically shifts to a higher level of information processing, as the consciousnesses of our Characters 2 and 3 come online to process new levels of input. In an instant, our sensory systems are flooded with bold stimulation including bright lights, loud sounds, and direct touch, all of which had been muted by the liquid environment of the womb. Considering that our brain's wiring is a product of both nature and nurture, stimulation now streaming full force through our sensory systems is initially perceived as chaos because those systems are not yet fully mature. Yet our brain is a masterful tool specifically gifted in its ability to create order from disorder and make sense from nonsense.

At the time we are born, we have no definition of our physical boundaries and our brain cells require stimulation from the external world in order to establish the circuits and future networks necessary for us to define and control our muscles. As a side note, this is why it is so important that infants not be tightly swaddled for hours on end, and instead be allowed to flail their limbs about. When we are infants, every random movement of our limbs sends information from the muscles crossing the joints up to our brain about our position in space. When we are born, we are merely a collection of cells with an undefined and unrefined consciousness. These arbitrary motor movements are critical for normal brain development and should be encouraged. Our brain learns quickly, and as we gain awareness of our physical boundaries, we also achieve a rudimentary control of our limbs.

It is also important to be aware that when we are born our brains are not a blank slate, as our genetic profile carries with it innate and instinctual wisdom. The chromosomes of our DNA are made up of the exact same four molecules as other mammals, which means that we are coded at the level of our genomic inheritance to share patterned responses and insights from our genetic predecessors. Just as an example, we humans share 99.4 percent[2] of our genetic code with the chimpanzee, and a part of that coding includes instinctual and protective insights.

---

[2] Derek E. Wildman et al., "Implications of natural selection in shaping 99.4% nonsynonymous DNA identity between humans and chimpanzees: Enlarging genus *Homo,*" *Proceedings of the National Academy of Sciences of the United States of America* vol. 100, no. 12 (June 10, 2003), 7181–7188. https://doi.org/10.1073/pnas.1232172100.

Brain Cells and Consciousness

As I noted earlier, we are feeling creatures who think rather than thinking creatures who feel. When we are born, the circuitry of our emotional Characters 2 and 3 are far better developed than the higher-thinking circuitry of our Characters 1 and 4. When our two emotional Characters 2 and 3 come fully online, our overall attention becomes focused on filtering the inundation of sensations streaming in from the external world. When we start processing information as our Characters 2 and 3, we become distracted away from the awareness and the more subtle and omnipotent consciousness of our Character 4.

Functionally, the cells of our Character 4 exist at the neuroanatomical junction between the experiential physical life of our Character 3 and the boundless consciousness of the universe. In other words, it is the Character 4 part of our brain that *is* a spiritual being having a physical experience. And as such our Character 4 is our connection to our Higher Power whereby we exist as a part of the Infinite Being. Use whatever language is comfortable for your belief system, but this character exists as the cosmic consciousness.

Inside the expanded consciousness of our Character 4 . . . we are perfect, whole, and beautiful.

With time, this consciousness of our Character 4 energy ball, which is both the life-force power of the universe and the consciousness of our cells, shifts silently into the background of our perception. The easiest way I know of to hook back into this eternal state of peacefulness is to consciously choose to bring my mind to the present moment and then expand my consciousness to embody a deep sense of gratitude. I do this regularly through the practice of the Brain Huddle, which we will focus on in the next chapter.

Inside the expanded consciousness of our Character 4, where we have no physical boundaries or sense of our own individuality, we perceive ourselves to be both as big as the universe and enveloped in the deep, eternal love of the cosmic flow. The *feeling* of the cosmos, the *sensation* of an all-pervading experience of deep inner peace and love, is available to us in life and is what we will return to in death. Within this awareness that we are safe regardless of our physical circumstance, we can exhale into this experience of deep inner peace and contentment, where we are perfect, whole, and beautiful. The road to true enlightenment is knowing that this eternal peace is our future, present, and past.

One of my favorite books on this subject is *Why God Won't Go Away* by Drs. Andrew Newberg and Eugene d'Aquili. These scientists authored the famous research in which they studied monks and Franciscan nuns by using a SPECT machine to determine what happens in the brain during meditation or prayer. What these scientists learned was that instead of finding an area in the brain that would light up when these folks felt a connection with the eternal, God, or feeling at One with the Universe, they instead found that the language and other left-brain centers became silent.

Character 4 in the World

Our left-brain Characters 1 and 2 interpret the space between any two *things* to mean that they are separate. Science teaches us that there is an electromagnetic field of atoms and molecules that surrounds us, and that we exist within it and it exists within us. Our left brain is unaware of this sea of energy because a small group of cells in our parietal region defines the boundaries of our body, surmising the separation. How different would our world be if we understood that we have the power through our thoughts and emotions to influence this field of energy? Perhaps gaining this insight into whole-brain living is humanity's collective Hero's Journey and how we will evolve as a species to live our lives with purpose.

We humans are energy beings that transmute one form of energy into another. For example, through our sensory systems we transform vibrational patterns into sound or vision, which is completely dependent on the structure and function of those specific neurons. We are energy beings having a physical experience, and we are not simply vibrational receivers or mechanical achievers through the use of our muscles and limbs. Instead we have the power to organize our thoughts and communicate via language using the vibrational tones of our voices, as well as many of the more subtle forms of communication that we discussed as gifts of our Character 3.

The energy that fuels the movements of the planets and stars is the exact same energy that forms the consciousnesses of both the overall universe and our Character 4. There is no division in the particulate matter that makes up everything, and it is all in motion. Because we are neither disconnected nor separate from the universal flow, we humans have the power to focus our minds and emotions and purposely shift that energy. Through the power of prayer and the power of setting our intention, we have the power to consciously change the way that energy flows.

When we purposely focus our intention and send those vibrations out into the great unknown, change happens. We are powerful beyond measure, not just within our brains but in how we can use our brains to influence the energy field and thus the world around us. At its core, this is why the book and movie *The Secret* had such mass appeal. Energetically this relationship between us and the space around us is real.

I hook into my Character 4's consciousness that exists in the eternal flow when I shift my mind into the present moment, focus on my breath, and feel my heart expand to connect with the breeze that both brushes my face and simultaneously rustles the leaves on the trees. When this happens, I shift out of my left brain's perceived boundaries and meld into the energy. I *become* the movement of the flow and shift into *being* that elusive thing. I am not only the leaf, but I am the energy that moves the leaf. I am not just the bird that soars, I am the energy that lifts the wing so the bird can pivot even higher. I am not just the kiss of the breeze upon my face, but I am the warmth within it. I am not just the purr of a kitten, but I am the energy of love that radiates in that vibration.

I connect with my Character 4 by extending myself out into the sparkling colorful light of a rainbow as I remember what it feels like to sparkle. I shift into my Character 4 as I devolve into the energy of that loving gaze between a mother and her nursing infant. My

Character 4 is everywhere and in everything, and during these moments I delight in the glorious bliss of just being alive. My Character 4 revels in the bond I've grown to know with the great blue heron as it squawks to me "good day" on its morning rounds, and at dusk I empathize with the owl as she calls for her mate to come home for dinner. Together in that shared and sacred consciousness, we are all family here as one ball of energy.

There are so many wonderful poets and musicians who have iterated the bliss of our Character 4: beautiful souls who have seen themselves in the reflection of God, not separate from it, and then shared that with the rest of us. The gift of poetry in the form of music fluidly seeps meaning into the distant crevasses of our being, and nothing has ever more completely summed up the soul of my Character 4 than singer/songwriter Carrie Newcomer's *Bare to the Bone*:

> *Here I am without a message*
>
> *Here I stand with empty hands*
>
> *Just a spirit tired of wandering like a stranger in this land*
>
> *Walking wide-eyed through this world is the only way I've known*
>
> *Wrapped in hope and good intentions and*
>
> *Bare to the bone*

Just as our Character 4 can profoundly influence our use of language to converse in deep and soul-stirring ways, this part of our consciousness is open, aware, and accepting that everything is exactly as it should be. Our Character 4 does not judge, it simply celebrates with wonder the life that it lives. This part of our character teaches our other characters that we are not just worthy of being loved but that we *are* love. When our left-brain characters open their hearts to the wholeness of their right-brain Character 4, any lack of worthiness instantly disappears. We cannot experience ourselves as unworthy of love and be the love of the universe in the same moment.

Because of the way our brain develops, children are often more comfortable with their Character 4s than adults. As we age and assign higher value to the skills and consciousnesses of our left-brain characters, we become more comfortable in the external reality of our left brain and less comfortable in what becomes unconscious and unknown. This makes sense, of course, because we learn the rules of society and how to focus at the level of *things* early on. We learn how to pick up our toys, how to not get lost in the grocery store, and how to use our *inside* voice. We are trained when we are very young how to follow the rules of society and honor the values of the world.

Consequently, for many of us, the mere idea of dropping our left brain's focus on identity feels like a terrifying death, but it's really more like wading into a cold mountain stream on a hot summer day. At first it feels shocking, and then you go a little deeper and your body begins to acclimate. Before you know it, the water is waist-high, and although this may be the coldest water in which you have ever been, your body tingles and your whole soul wells up with amazement as you dive in. Taking that first step into the water is heeding the call of your Hero's Journey. Eventually your ego acclimates to the understanding that it will not die if you simply set it momentarily aside. Your ego remains right there, always available to be immediately brought back online if you choose that. But when you permit yourself to embark upon the journey, both insight and growth await you.

It can be simple for us to step into this consciousness of the unknown when we strip off our judgments, our schedule, and our worries and choose to really enter the realm of the present moment. Go jump into a mud puddle without thinking about the mess or consequences, and encourage that overwhelming sense of jubilee to erupt from your soul. Remember that childlike joy you used to experience? Just thinking about it radiates a smile right down to my core. When we are willing to step out of the properness of our suit, our ego, or our justification of our value, and we let ourselves land in the here and now, life is messy, so let it be what it is and revel in it.

Who are you when you set down your self-doubt, judgment, and criticism? Who might you be if you believed in yourself like your Character 4 believes in you? Who might you be if you identified and embodied this part of you in every moment of every day? How vast would you be if you set yourself free from those left-brain boundaries and limitations? This Character 4 part of you is always right there, connected to everything and loving you perpetually. It is the majesty of the mountains when we pause and breathe it in, and it is in the energy of the ripples as they dance across a surface.

This sacred awareness of your Character 4 is omnipotent and just beyond your focus, and no one else can take you there but you. Although you may be physically alone, this part of you is incapable of loneliness because it is the love that is woven into the consciousness of everything. Our Character 4 is grateful for the gift of life, accepts what is, and delights in the passing of time.

As the poet Rumi so eloquently invited, "Out beyond ideas of wrongdoing and rightdoing, there is a field. I'll meet you there. When the soul lies down in that grass, the world is too full to talk about." Our Character 4 is this part of who we are that resides both at the core of our being and at the edge of the veil of whatever lies beyond. I will meet you in that field . . .

CHARACTER 4 AT WORK AND AT PLAY

Character 4 on the Job

Character 4 is the anchor of any corporate ship. Each of the other three characters rocks the boat in their own unique ways, but Character 4 is the predictable, reasonable, big-picture, unbiased perspective of how it all fits together, flows collectively, and works—or not. Character 4 is not afraid of the money situation, no matter what it is, and is not attached to its ego-center because it does not have one. It is aware of the egos of others, of course, but Character 4 is completely available to assess the big-picture performance of the overall machine. It thinks in systems: *If we do this, then that will happen, so then we need to counter that with this in order to create balance.*

Our Character 4 can juggle nine details at the same time and not be paralyzed with fear or feel overwhelmed by a complex task. Character 4 looks at all the parts as different pieces of the whole, but its strengths and insights are in putting them together to create a systemic flow. As a result, our Character 4 is the beta test of any organization. If I don't know whether something is going to work, I can take the idea to a Character 4, who will visualize the project and then tell me *yes*, we can do that or *no*, that will not work. The Character 4 may also say, *yes*, we could do that, but I'm not sure that we should. If the Character 4 is not convinced that the gain or addition will add value to the bigger picture, then it will always choose simplicity and clarity over complexity.

When it comes to business practices, our Character 1 wants to make a profit, our Character 2 will squirrel around with the idea and details, our Character 3 wants it to be fun, and our Character 4 wants to serve the greater good.

Character 4 at the Beach

Character 4 can hear the sounds of the beach, the breaking waves and the birds calling long before they get there. Character 4's heart is so filled with gratitude as they connect with the vast expansiveness of the ocean that ideas of hope and possibility immediately replace any inklings of despair. In this space, in this feeling of connection with something greater than oneself, there is a sense of total abundance and total surrender to an all-knowing awareness that everything is exactly as it should be.

We may be alone at the beach, but our Character 4 is never lonely. We feel innately at home and completely present as we connect with all that is. We watch the birds for longer than we realize as we get lost in their musings. The essence of our being ebbs and flows in rhythm with the waves, and we soar with the birds and exude contentment. As the sun warms our skin, we close our eyes, raising our arms to the heavens in appreciation for our life and all the life around us. We exhale.

Consumed by a feeling of grace, we know at our deepest core that we are perfect, whole, and beautiful just the way we are. We make no comparisons because we are completely here in the present, with a mind that is void of another place or time. We are grateful that we have life, that we are life, and that we share life. We feel humor in the playful dance of the pelicans, we see meaning in the clouds, and we feel the beauty of the magic all around us. We are aware of the tapestry of which we are a part as we move within the flow of it all.

Character 4s at the beach smile easily and share direct eye contact with those passing by. They are in motion energetically, even if their bodies are not in action. They connect in amusement with the wild screams of children at play and smile at the old geezer who is napping. If you open your soul to their wavelength, you can feel their projection of love to everyone and everything. And those dolphins we see today, the ones that bring everyone so much joy, they are here in communion with our Character 4s.

A Snapshot of Our Character 4

- Aware: *I am connected to all that is. I am aware that I share the same consciousness as everything around me and that I am in it, and it is in me. We influence one another even though we cannot see it. We can train ourselves to feel it and know it.*

- Expansive: *I am open to possibilities and value the big picture and wholeness of my existence. I do not fear the absence of my ego-self because I know that I am perfect, whole, and beautiful just the way I am. We exist in the energy of the One.*

- Connection: *In the consciousness of the cosmic flow I embrace the timeless, all-knowing part of myself that is connected to all that is. I resonate in this space when I meditate or pray. Each one of us is a neuron in the network of humanity, and we are intricately connected at the molecular level as a part of the flow.*

- Accepting: *I can either accept life on its own terms and feel peace knowing that everything is as it should be, or I can be attached to how I want things to be and suffer when reality does not match my preconceived vision.*

- Embraces Change: *I love and celebrate what is right here in front of me, and then when this moment, this life, this love, this experience is past, I am grateful that I had it at all. Life is a series of moment-by-moment changes, and I embrace them all with an open heart and am grateful for whatever comes next.*

- Authentic: *When I strip away the details of who I am in the external world, when I step beyond the details of my Character 1 Persona, Character 2 Shadow, and even my Character 3 Anima/Animus, I own my power and step forward as my best self because this is the consciousness of the Higher Power that flows through me.*

- Generous of Spirit: *I am a part of the whole. When I give to you, I am giving to a part of myself. When I help you, I give assistance to us all. When I love you, I accept you as you are and we all thrive.*

- Clarity: *No longer distracted by the workings of the external world, I am clear in our intention to love and be loved. Our number one job in life is to love one another. Period.*

- Intention: *I set my intention and trust that everything is connected and in flow. When I use the power of my heart and mind to manifest something, I am using the power of myself to shift the arrangement of the atoms and molecules in space, and I stay on course.*

- Vulnerability: *Stepping into life as the consciousness of the universe, I can step into the nakedness of my Character 4 and stand strong in my vulnerability. When I let you see who I am, I empower you to do the same.*

Getting to Know Your Character 4

As with all the other characters, if you prefer to skip these questions for now, feel free to continue reading, and come back for this exploration when you can take some time to really explore this part of yourself.

1. Do you recognize your Character 4? Pause for a moment and imagine yourself being this part of yourself.

I became really familiar with this part of my being on the morning of the stroke, when I shifted completely out of my left-brain Characters 1 and 2. I felt as though there was nothing left of me, the individual I had known, yet I was still tethered to this body and this life. In the absence of everything that I had grown to be, I felt pure bliss and grace in the all-knowing consciousness of the universe. I knew that although I was not yet dead, I was as disconnected as I could possibly be and still be counted among the living.

2. What does Character 4 feel like inside of your body? How does this character hold your body, and what does your voice sound like?

In the absence of my left-brain characters, the consciousness of my Character 4 was one of uncompromised peacefulness and euphoria, and I purposely return to that state many times every day. When I am embodying the consciousness of my Character 4, my vision blurs, my senses perk up, the feeling in my chest expands, I focus my attention in the present moment, the boundaries and edges of my body slip out beyond my awareness, and the essence of my being swells in a timelessness of intense grace and contentment. When I speak, my voice drops into a lower register and I enunciate clearly. The consciousness of my Character 4 is an amazingly beautiful awareness that I know one day I will return to full-time, and when I do it will be a true coming home.

3. What if you don't recognize this character inside you?

If this Character 4 feels completely unfamiliar to you, and you not only cannot recognize this part of yourself but it sounds completely absurd, foreign, or perhaps even dangerous, rest assured that you are not alone. Just as with our Character 3, it can feel very threatening to our left-brain Characters 1 and 2 to step out beyond our sacred individuality. However, offering techniques and tools that are designed to help people find this place of deep inner peace is a thriving multibillion-dollar industry, and if you choose to use tools to help you get there, you have many options from which to pick.

Because we live in a society that is skewed toward the hierarchical, materialistic values of our left brain, which focuses on me, the individual, we are rewarded for what we do rather than for who we are. Our right-brain Character 4 exists simply as life in a peaceful and blissful state of feeling connected to all that is, whereby life itself is the reward and gratitude is the underlying feeling. It is impossible for our Character 1 to linearly rationalize or think its way into our Character 4. Instead we must surrender to it, and that can feel scary.

If your left-brain Characters 1 and 2 are really powerful, it may be difficult for you to step into your Character 4, because in order to do so you must feel safe. Not just physically safe, but safe emotionally from the judgment and criticism of others who don't value their own Character 4. Many strong left-brain characters will negatively judge the value of their own right brain, as they only believe in the reality that is defined by their five senses. In addition, few things stir as much fear, antagonism, or argument as the subjects of religion, spirituality, and other intangible forms of belief. We have reasons for believing what we believe, and any challenge to what we believe is often interpreted as a personal endangerment.

Having said that, the dogma and stories associated with religions are a function of our left-brain language centers, while our experience of spirituality and our connection with a Higher Power happen in our right brain. Regardless of which religion or practice you might observe, the ultimate goal of prayer, mantra, or meditation (at the neuroanatomical level) is to slip our consciousness out of the confines of our left brain, where we experience ourselves to be individuals, and into our right-brain consciousness of our Character 4, where we experience a fluid connection with our Infinite Being.

If you are a nonbeliever, either an atheist or agnostic, you are not alone. Many of us don't know what to believe, so we choose to not believe in anything at all except perhaps the power of ourselves and our five senses. Yet regardless of our level or subject of belief, when it comes to our brain, tuning in to the consciousness of our Character 4 can be very healing physically, emotionally, and spiritually. When we choose to team up with the consciousness of the universe for healing, amazing recovery can happen. Just look at my brain. I can guarantee you that it was not the power of my left-brain characters that brought them back online, it was the power of the universe that worked with the consciousness of my Character 4 to help my cells heal.

If you still cannot identify with your Character 4, consider the moments in your life when you have felt your heart expand or open. For many, the sighting of a rainbow, or fireflies, or even catching a glimpse of the shadow of a leprechaun as he dashes behind a tree at dusk might bring a special thrill to your soul. If you are not aware of ever having any of these feelings, there are tools you can use to help yourself become more aware, more open, and more connected to the all-knowing. If you are willing to expand your experience and practice shifting your attention toward the elusive, it is possible for you to train yourself to "become" the energy around you.

4. Assuming you can identify your right-brain thinking Character 4, how do you let this character express itself? How much time do you spend in your Character 4, and what is that like for you?

This part of me is never far away as I vibrate in the hum of the hummingbirds and coo at the trail of a shooting star, not forgetting to silently make a wish . . . *why not?* Character 4 is always running just below the radar of the noise that is made by my other characters, and this is my authentic self that is all-knowing, all-loving, and connected to all that is.

This portion of my brain knows that we are perfect, whole, and beautiful regardless of what is spiraling around me in this world or what circumstances I find myself in. And a dead giveaway that I am in my Character 4 is when I burst into song with Doris Day singing, "Que Sera, Sera." For you it might be "Don't Worry, Be Happy" with

Bobby McFerrin or "Hakuna Matata" from *The Lion King*. Regardless of our age and era, these songs sing the promise of our Character 4.

I step purposefully into my Character 4 when I exhale and allow myself to be held by something that is greater than myself. God is in the exhale. On the inhale are our expectations, our must-dos, our performance, our self-reliance, our self-judgment, and our anxiety. But when we let that go, when we surrender our attachment to what we want as opposed to what is, our Character 4 steps out and celebrates.

With practice, I shift my consciousness out of my normal focus, pushing the constant noise of the world into the background, and bring my attention to that which has no name. When I focus on my breath, I immediately shift my awareness to the present moment. When I think about my breath, I shift out of the past or the future. Once in the present moment, I connect with my sense of gratitude and allow my consciousness to devolve into the babble of the babbling brook and empathically experience the laughter, tears, and fears of others.

I let myself expand, weaving my energy between the blades of wheat in that field over there, and I am in the movement of the grass and trees over here. Matthew 6:28 says, "Consider the lilies of the field, how they grow. They don't toil, neither do they spin." They simply are, and they trust that is exactly what they are meant to be and do.

Our Character 4 is our mystic self. It is our knowing that we are not only held by nature but we are the crescendo of the insects as they perform their symphonic cacophony. We are the sunbeams when they explode as glory through the clouds, and it is my Character 4 that reads the Morse code in the surface ripples when they scatter across the water as a love note from our beloveds who are now out and beyond.

5. As you think about your right-brain thinking Character 4, can you assign it an appropriate name?

I lightheartedly and enthusiastically call my Character 4 "Queen Toad." I refer to her as a queen because, well, she is a queen. She is the part of me that is regal and connected to the almighty. I call her Toad because I am a goofball and I live on a lily pad on the water, a boat named *BrainWaves* where I spend five months a year. I

have learned that it is important for me to not take myself too seriously, since I am both the center of the universe and simply a speck
of stardust. My Character 4 is more than my life. She is the part of
me that is omnipresent, and she feels like eternal Love.

6. Who are some of the Character 4s who have influenced you over the course of your life, in positive or negative ways? Was your Character 4 emboldened by their Character 4 or repressed by it?

Following the stroke, I did not want to give up this awareness
that we are truly perfect, whole, and beautiful just the way we are.
So I made a vow to myself that I would only recover as much as I
had to recover for the rest of humanity to perceive me as normal.
By definition, the price I paid to be fully human again was losing
my complete connection with the Infinite Being. I made the decision to recover, however, because there was no point in me having
this experience with stroke and God if I was not going to come
back and share it.

I am often asked if I can return to that space of my Character 4 at
will, and most of the time, instead of living here and visiting there, I
choose to live there and visit here. I am a Character 4, and I add the
skill sets and circuitry of my other Characters, 3, 2, and 1, to function
here in this world as a living being. Yet I remain clear that Queen
Toad is, in Jung's language, my *Self,* and the rest of my consciousnesses are simply the other characters I use to live this human life.

I love meeting other Character 4s. They are truly the icing on
this cake of life. Put a couple of Character 4s together and watch the
sparks fly, the lightning strike, and the love explode. We are divine.

7. Who in your life appreciates, cares for, identifies with, and wants to hang out with your Character 4? What are those relationships like?

Some of the greatest words of wisdom I have ever heard came
from my very good friend Dr. Jerry Jesseph when he stated, "We, as
humanity, are confused and more confused." I think it is safe to say
that those of us who are most confused are completely unaware of,
disrespectful of, or do not value our or anyone else's Character 4.

Having said that, many of my closest friends live the life of
a strong Character 4. These friendships are all accepting, deeply

loving, supportive, nurturing, empathic, and kind. Character 4s share a knowing when they meet, and they are some of the most gratifying relationships one will ever find.

Who else likes this part of me? My fellow travelers who seek or know this part of themselves. Those who exist in that realm dance with me among the stars and know there is no separation in space and time, so we never need to meet in this life. Yet when we do, it is a sacred moment of connection and a shared moment of knowing.

8. Who in your life does not get along with your Character 4?

This part of me is pure love and sees the beauty in you no matter your human circumstances. This is the part of a mother who unconditionally loves her child as well as the children of others. This is the part of ourselves that knows no stranger, and the part of the Christ in us when we state, "Forgive them, Father, for they know not what they do." Character 4s love unconditionally, and at all times.

Our Character 4 is a safe place because it is compassionate, kind, and open. When someone else is having a bad day, my Character 4 can be kind and supportive and I can offer you a smile and an appropriate touch, even when you are growling. A Character 2 in distress can feel the strength, courage, and love when a Character 4 holds the space for them. Our Character 4 is the most powerful tool we have to de-escalate either our or another's pain.

9. What kind of parent, partner, or friend is your Character 4?

It is important for us to help our children develop a strong Character 4 because this is a meaningful healing part of ourselves. More than anything, healthy children want real connection with others, and as their parents or friends we can model for them what it means to have a healthy connection with our Higher Power. When we recognize and value the thread of the One that is in everyone and everything, we open our hearts and soften the edge of our judgment. We have the ability to invite our children into an awareness of the present moment, but more than likely they will probably lead you into that conversation since we are born in the consciousness of our Character 4.

One of the greatest gifts my mother ever gave me when I was a child was to assure me that even in those moments when I upset her, during those times when she may not have liked my behavior very much, she always loved me. In addition, G. G. was always on a path of growth, and throughout my life she let me grow and change and did not limit me to who I was yesterday. She let me grow out of bad behavior and did not hold me back.

When I graduated from college, I drove from Indiana to California to become a guide on the American River. A fantastic woman who was about my physical size volunteered to train me to row the rapids. Ragina, who was a 13-year veteran guide at the time, taught me that because of my size, I had to learn to row with my brain rather than my back. The men could muscle their way out of trouble, but smaller women could not. I grew up that summer on the river by meeting the best part of myself, and the part of me that I wanted to become. I met my Character 4.

When I returned home, my mother recognized that I was now a different woman, and she never held me back to my smaller self. By the time I had my stroke, my mother knew my Character 4, and she completely trusted that her job was to set me up for success by working with me to help me heal myself. My mother was my first and greatest blessing, as she reared me twice.

By modeling for others how we find holiness and sacredness in the simple things in nature, we help others see it more clearly. I had a friend who would always pick worms up off the sidewalk and toss them back into the yard when we went on walks. This was a great practice, of course, except on those rainy days when thousands of worms were unearthed. Then I found my left brain questioning the wisdom of what we were doing, while my right brain thought maybe it was time to go fishing.

10. How kind is the relationship inside your head between your characters? How does your Character 4 relate to your other characters?

Queen Toad does not just love everyone, she is love. She respects and supports my Character 1, Helen, and celebrates her efforts to help our life have order. On cue my Queen Toad zips right in to hold and nurture my little Character 2, Abby, when she stumbles into fear or pain. Queen Toad adores my Character

3, Pigpen, but has to remind her often that although we are fine with death and dying, we would appreciate Pigpen's cooperation in helping us keep ourselves alive. So far Pigpen has accommodated this request, and we come in out of the thunderous lightning earlier than we used to.

Cheat Sheet: Get to Know Your Character 4

1. Do you recognize your Character 4? Pause for a moment and imagine yourself being this part of yourself.

_____

_____

2. What does Character 4 feel like inside of your body? How does this character hold your body, and what does your voice sound like?

_____

_____

3. What if you don't recognize this character inside you?

_____

_____

4. Assuming you can identify your right-brain thinking Character 4, how do you let this character express itself? How much time do you spend in your Character 4, and what is that like for you?

_____

_____

5. As you think about your right-brain thinking Character 4, can you assign it an appropriate name?

_____

_____

6. Who are some of the Character 4s who have influenced you over the course of your life, in positive or negative ways? Was your Character 4 emboldened by their Character 4 or repressed by it?

_____

_____

7. Who in your life appreciates, cares for, identifies with, and wants to hang out with your Character 4? What are those relationships like?

_____

_____

8. Who in your life does not get along with your Character 4?

_____

_____

9. What kind of parent, partner, or friend is your Character 4?

_____

_____

10. How kind is the relationship inside your head between your characters? How does your Character 4 relate to your other characters?

_____

_____

CHAPTER 8

The Brain Huddle: Your Power Tool for Peace

I love watching my Four Characters interact with one another as they collectively live my life. From moment to moment, my Four Characters flit in and out of dominance, as I really do independently embody all four of them. I might be onstage giving an interview with Character 1, Helen, teaching about the brain, and then in an instant Character 3, Pigpen, may leap into my consciousness, take over the mic, and give an example using my full body as a prop. Trust me: Helen would never have done it that way. But now, instead of feeling shock or embarrassment, Helen has learned to appreciate Pigpen, and she actually finds her as entertaining and brilliant as the audience usually does. No one can come up with a better experiential example to make that point that Helen is dryly teaching than Pigpen. The more familiar you become with who is who inside of you, and the safer you feel to exhibit each of them, the more wholebrained your life will become.

Having watched these Four Characters over the last few years, I have seen that much of their behavior is relatively predictable. Helen likes to hang out in the office or on the phone, while Queen

Toad inevitably shows up when I am on a nature walk. At least until something really fun happens, and then Pigpen jumps in for an adrenaline rush. Each of our Four Characters is somewhat stereotypical in their behavioral patterns, so the more willing you are to observe yours in the wild, the more freedom you (and they) will feel to live a whole-brain life.

Creating healthy relationships between our own Four Characters and the Four Characters of others is our goal, especially if we would like for most of those interactions to be positive and life giving. We do have the power to pick and choose moment by moment which of our Four Characters we want to be in the world, and sending them into a Brain Huddle is our key to our next best action.

Let's take a deeper dive into what it really means to take a Brain Huddle as we gather together the voices of our Four Characters into a conscious conversation about what is going on in any moment of our life. The Brain Huddle is how we own our power by taking full responsibility for who and how we present ourselves to the world, as well as how we choose to let the world influence our thinking, emotions, feeling, and behavior.

When we run our lives on automatic, our Four Characters do whatever they please without any real consideration for what we might actually *choose* to be doing instead. When our Four Characters gather together in a Brain Huddle, just as in a sport, they each share their perspective and then collectively choose our best next strategic play. Regardless of what is going on outside of ourselves, we have the power to routinely come out of the huddle with an appropriate and peaceful resolution.

By choosing to engage in a Brain Huddle, we set ourselves up for success. If we really want to live our lives on purpose, let's do it both systematically and enthusiastically. Besides, once we call a Brain Huddle and our Character 4 is engaged in the conversation, we are pretty much guaranteed a loving outcome.

An Anchor for the Heart

Just as with the other neural circuits in our brain, the more time we spend practicing the Brain Huddle during the benign moments of our lives, the stronger that circuitry becomes. Eventually our capacity to take a Brain Huddle will run on automatic as a new habit. Imagine how different your life will be when taking a Brain Huddle happens spontaneously throughout the day, and it becomes so practiced that it is now your brain's normal response when you feel your emotional triggers are being pushed.

By far one of the greatest gifts of knowing my Four Characters, and perfecting the ease with which they collectively enter into a Brain Huddle, has been the realization that when I am feeling isolated and caught up in the drama or emotional pain of my Character 2, that part of me is not alone. When Abby is caught in the loop of feeling sadness or hopelessness, as soon as we call a Brain Huddle, any feeling of isolation or despair immediately dissipates. It is impossible for my right-brain Characters 3 and 4 to feel emotional loneliness, since they exist in the consciousness of the collective whole.

However, during those moments when my Character 2, Abby, does feel disconnected from both the rational brain of my Character 1 and the collective consciousnesses of my right-brain Characters 3 and 4, she can feel totally engulfed by a blinding fog of desperate emotion. When experiencing this level of duress, just being aware that my Four Characters are there and that I can call on them to hold a Brain Huddle has been an anchor for my heart.

I can attest that nurturing the habituation of our Brain Huddle will lead to a more whole-brain life. Every once in a great while, my Character 2, Abby, experiences a debilitating anxiety attack. I never know ahead of time when this is going to happen, but when it hits I feel like my brain is caught up in an internal tornado of emotional reactivity. Having the option of shifting my Four Characters into a Brain Huddle during those moments of need has been a real lifesaver. Now, when my little Character 2's emotional stress circuitry even begins to rev up, I have a well-developed tool I can use to save my peace of mind and calm my physiology.

If you, too, experience anxiety, you know how running that circuit has the power to completely hijack your brain. Raw fear, anxiety, or even a panic attack: each has the power to render us emotionally incapacitated, feeling scared, desperate, vulnerable, and isolated. In those moments when I feel completely enveloped by fear, I have found comfort in just knowing that my other characters are still there. Even though I may have lost all sense of them, I know that they are there watching, waiting for this condensed energy surge to spread out to include them again so they can engage again. I am sure that this tool of the Brain Huddle will help you, during your time of need, in the same way that it has helped me survive those dark moments of my soul.

In those moments when I feel completely enveloped by fear, I have found comfort in just knowing that my other characters are still there.

If you don't have any personal experience with severe anxiety or know what it feels like, imagine all the energy in both your brain and body suddenly flushing up into the group of cells making up your fight-flight limbic circuitry. Your ears hear the roar of your blood as it sounds the alarm of your heart pumping so loudly that you cannot think clearly or see straight. You are thrown off balance as your entire body surges in what feels like a violent physiological attack, rendering your Character 2 completely flushed and overwhelmed with vulnerability. Remember how we are feeling creatures who think, rather than thinking creatures who feel? When that alarm of our emotional Character 2 is triggered in a full-force fear, anger, or anxiety attack, just knowing our other characters are right there, ready for a huddle, is a lifeline.

Another wonderful gift that the habit of the Brain Huddle has brought to my life is that I have used this tool to train myself to easily identify and embody each of my Four Characters at will. Now, when I am feeling even the subtle invitation of my Character 2's emotional reactivity tugging on my consciousness, just knowing intellectually that my other characters are there somewhere beyond the disconnect has allowed me to shift my focus away from engaging with my Character 2 to simply observing her.

At an energetic level, as soon as I make this perceptual modification of observing instead of engaging, the energy ball that is focused on stimulating that emotional circuit begins to dissipate. As that tight focus of energy starts to expand, it reinfiltrates the other parts of my brain, and within moments my other characters can come back online and reenter my consciousness. As they reappear and join Character 2 in a Brain Huddle, Character 1 steps up to make sure we are physically safe in the immediate moment; Character 3 starts running imaginary scenarios about how we might strategize a successful play; and Character 4 holds the space for us all—realizing that even in the worst-case scenario, no matter what happens, we are okay.

Our Mighty 50 Trillion

When I experienced the stroke and my left brain went off-line, I pictured the 50 trillion cells making up my body and brain as beautiful little creatures, each with a consciousness of their own. I listened to them. I communed with them. I valued their efforts and shared with them how I needed them to help me heal, and then I encouraged their efforts. I still give credit to my brain cells for my recovery, and I continue to honor their ongoing efforts to heal my wounds day by day.

I believe our cells have the power to heal us. Of course I value traditional medicine, especially in the case of an emergency, but I do believe that a little faith in our own power to heal creates a respectful and healthy team spirit among our cells. As a collective of 50 trillion molecular geniuses, we share the common consciousness of our Character 4. When we are willing to tap into the combined power of the consciousness of the universe and our cellular Character 4, we heal.

When my brain was devastated, the consciousness of my Character 4 was all that remained. Once I decided that I was willing to *try* to recover, which meant that I was willing to endure the hard work and agony of trying to focus my attention on things in the external world, my Character 4 embarked upon the journey by taking total responsibility to love and direct all the healing efforts of all of my cells. As a collective consciousness, all the cells of my brain and body worked together, and although I had no idea in the long run how much I would heal or recover, my cells and I joined our forces by collectively tapping into the conscious power of the universe.

Having recovered completely, I define and value each of my Four Characters in a similar way. I see them as individuals, yet they are each a part of my whole brain. I listen to their needs and call on them accordingly. In many ways my Four Characters are like four children who value being heard and seen, and getting to know each of them has made the little moments of my life more predictable, for both me and those around me. My Character 1, for example, is going to answer the phone on the first ring, but if a friend calls and it goes straight to voice mail, they can rest assured that my Character 3's hands are busy.

Each of my Four Characters has merit, but as individuals they live by their own set of values. Character 1 will dress well and arrive promptly at an evening engagement, while Character 4 may choose to miss the event altogether so she can stay out and commune with the sunset. Unless I am on a strict schedule, I let my Four Characters lead my life. Am I feeling refueled and ready to go out and share my energy as Helen, or do I prefer to refuel and harmonize with nature as Queen Toad? Imagine the authentic peace we would each bring to the world and one another if we completely supported our own, as well as each other's, Four Characters.

Benefits of the Brain Huddle

In order for us to make good decisions, we need to know what our options are. Before I thoroughly understood the Four Characters material, I didn't really know how to choose anything other than the black-and-white, obvious alternatives. Although I usually feel good about the decisions I make, every now and again I (probably Pigpen or Abby) make a decision and then realize that if I had paused and implemented the Brain Huddle, I would have chosen more wisely. I have found that the Brain Huddle soothes my insecurities and gives rise to the voice of my most authentic self, which is truly the combination of all of my voices.

There are several benefits to using this tool. First, the Brain Huddle requires that I push the pause button, which is essentially the same as running the 90 Second Rule that I described in Part I. Pausing for 90 seconds allows whatever chemicals are running through my bloodstream to flood through and then completely flush out of me. Once I am clear-minded again, and no longer feeling whatever I was feeling, I can bring all Four Characters into the conversation and make better decisions.

Second, the Brain Huddle encourages all Four Characters to voice their opinions. I run my brain like a democracy where every character gets an equal vote—unless I am in danger. When each character feels heard, and they listen to the opinions, wants, needs, and ideas of the other characters, when they come to a consensus it is a unanimous decision.

As a result, the third clear benefit of the Brain Huddle is that any decision I make using this tool is fortified by the support and consensus of all four of my characters. When I own my power in this way, I am confident that I have made the best choice. As a consequence, my most authentic life is supported by this tool that promotes healthy whole-brain living.

But equally important, because I understand the Four Characters—who they are and what they value—I get to see how the Four Characters are playing out in the lives of those around me. My awareness of who you are gives me a terrific advantage in how I might choose to interact with you. It does not give me an advantage over you, but it certainly offers me insight into how I might support and interact with you most effectively, since I always want to have a more peaceful conversation, negotiation, and resolution.

Trying to understand another person's point of view sets us up for clear communication. Loving people as they are, and not needing them to change for us to feel safe, is also a gift. There are eight characters in every relationship, each with individual needs, opinions, and desires. Recognizing another person's Four Characters and their needs can offer us a road map to achieving peace and harmony during our communications. If I find that I am experiencing conflict with you, for example, this tool helps me step back and assess my response in a calm way, rather than reactively. I can then evaluate the situation and show up more compassionately to listen to you.

The most important reason why I use the Brain Huddle on a regular basis is because it is a road map to my best self.

The Brain Huddle has proven to be a fantastic power tool for quick and precise communication. For example, the other day I called a friend who is a primary Character 4. When she answered the phone and we started chatting, I said, "Wait, before we get to your four, tell me, how is your one?" She proceeded to share with me her Character 1's projects and how those were going. Then I asked about her Character 2 and she shared with me that her 2 was feeling particularly tender because of an encounter she had the day before with a family member who was not well. Then she burst out with what adventures her Character 3 was knee-deep in, and then we touched home about her Character 4. Within five minutes my friend and I had an amazingly meaningful and connecting catch-up about all eight of our characters. It was fun, clarifying, and deeply satisfying. We were both grateful for the insights of the Four Characters language.

Using the Brain Huddle for personal reflection can help us determine what changes we might like to make in who and how we present ourselves to be, as well as in how we choose to connect with others. I realized that every time I drove over to visit with my mom, G. G., it was essential that I show up as my Character 3 rather than as my Character 1. G. G. wanted to set the schedule and make the decisions about what we did, so if I wanted to keep the peace, on the road over to her place I would consciously quiet Helen and wire up

Pigpen. This was the magical combination that worked for our relationship. I'm sure you have some of those relationships in your life too. Knowing how to navigate our relationships in this way is part of how we can choose to love one another.

The overall most important reason why I use the Brain Huddle on a regular basis, however, is because it is a road map to my best self. This tool allows my two left-brain ego-selves to have a voice and be heard, but in the end, the voice I really want to bring into the conversation is the voice of my unconditionally loving Character 4. When I enter into a Brain Huddle, I know that I won't come out of it with a decision until all Four Characters have weighed in. Predictably, I know that as soon as my Character 4, Queen Toad, shows up and participates, my Characters 1, 2, and 3 can relax and I will choose to take my best and most loving steps forward.

How It Works: The Conversation inside Your Head

The action of taking a Brain Huddle can serve as a balm for your whole being. As I noted in Chapter 3, the process of consciously and deliberately bringing all Four Characters into the conversation is both powerful and empowering. I have also found it to be grounding and deeply comforting. I like to think of it as a power tool for peace.

Here, I will outline the steps that make up a successful Brain Huddle and describe in detail the power that each of the stages brings us. As we saw earlier, the steps of the Brain Huddle spell out the acronym B-R-A-I-N, giving us an easy reminder, even during our most debilitating bouts of fear or anxiety. I can rely on this tool to help me recalibrate my brain when I am feeling emotionally reactive, overwhelmed, or vulnerable. I hope you find this tool to be as effective as I have.

B = BREATHE

BREATH is how I hit the pause button and bring my mind to the present moment.

At a neuroanatomical level, the power of our human brain is not in the way our cells excite one another but in the way they inhibit one another. It is really easy to get all cranked up, excited, and out of control, but it takes maturity to not be reactive and to stop running our circuits on automatic. Our willingness to push the pause button is how we engage the power of the Brain Huddle.

The best way I know how to create a physiological pause is to bring my attention to the present moment and focus on my body and my **BREATH**. Once my awareness is grounded in the right here, right now, I can sidestep the habituated patterns of my thinking and emotional circuitry that have been running on automatic. It does not matter which of the Four Characters I have been embodying. When I pause and focus on my **BREATH**, I break the circuit and create space for something new.

By choosing to focus on my **BREATH**, I am pushing the pause button between the stimuli that are streaming in through my sensory systems and my automatic responses to that stimulation, be they thoughts, emotions, or actions. Just as there is a space between the neurons making up a circuit, I have the power to stop cellular communication instantly. I do not have to run outdated patterned responses that are wired in my brain from the past. I can make new choices and consciously build new circuitry. Practice does make perfect when it comes to brain circuitry, because every time we run a circuit by choice, it becomes stronger.

As I focus on my **BREATH**, I connect safely and intimately with my whole being. I trust my **BREATH** implicitly, as it has always been with me like a comforting teddy bear, and it will stay with me for as long as I am alive. To the consciousness of my Character 4, the universe is the womb within which I am breathing, and I perceive that I am being breathed by the omnipotent consciousness of the eternal flow. The universe is breathing me, and I have life because it is there supporting my life. By thinking about my breath in this way, I feel my consciousness expand, and this is particularly helpful when I am experiencing a powerful bout of hostility or anxiety.

If I have been running my Character 1 skill set before jumping into a Brain Huddle, after a brief **BREATH** and pause I may choose to step right back into that consciousness, now feeling a little more relaxed. This part of my brain is masterful at running the details of my life, so stepping in and out of her with ease, and periodically taking a refreshing break, can be healing for my overall physical and mental health. It is well known that getting some movement, catching a power nap, or indulging in a distraction that breaks the continuity of a left-brain task encourages the brain at a neurological level to disrupt its stress circuitry, hit a reset, feel refreshed, and be open to new insights and possibilities. This is particularly effective when the left brain is struggling to solve a problem or write an intelligible paragraph.

When I focus on my **BREATH** and purposely bring my awareness of my physical body to the forefront of my mind, I tend to shift completely into my Character 4, whereby I feel a vast sense of gratitude that I am this miraculous being of life. At this level of awareness, I use my **BREATH** to consciously neuro-regulate my autonomic nervous system. Through the eyes of an anatomist, I can't help but recognize the similarity between the semipermeable membrane in my lungs that filters oxygen out of the air and the semipermeable membrane of that single cell that is attracted toward some things and repelled by others. There is a very fine line between this life and the absence of it. Breath is the key.

R = RECOGNIZE

RECOGNIZE which of the Four Characters are in the room.

Right here, right now, I have the ability to pause and breathe deeply, focus my mind on what it feels like to be me, and then **RECOGNIZE** which character circuitry I have been running and which characters of other people are in the room. I may also **RECOGNIZE** that all four of my characters are stammering for the microphone because they all have something important to contribute.

I **RECOGNIZE** that I am in my Character 1 when I feel like I am focused on details and concentrating on something, gathering information, organizing that information, or working methodically toward an end goal. As a Character 1, I am highly disciplined and enjoy being the boss. I thrive when allowed to be in control of myself, a situation, or others. There is a right and wrong way of doing things, so I am particularly good at establishing efficient systems and then perfecting the performance of those systems as time goes along. I am known to be precise, efficient, and competent, so I feel a personal sense of satisfaction when I do things well.

As a Character 1, I think linearly, and when I have a project to complete that is made up of many steps, I start at the logical beginning. I am a natural taskmaster, so fixing problems suits me well. It is easy for me to **RECOGNIZE** when I am running the *get on it and get it done* circuitry of my Character 1, and it is easy for me to **RECOGNIZE** this character in others.

But if I am feeling hurt, lonely, abandoned, or emotionally charged about something that has happened in the past, or an old hostility or injustice keeps rattling around in my brain, it is simple for me to **RECOGNIZE** that my Character 2 has been triggered. Anytime I am on fire with anxiety about the future, or totally freaking out that I am not being seen, heard, or treated fairly based on my old wounds, it is easy for me to **RECOGNIZE** that my Character 2 has flared right up in an attempt to not only protect me but help me get my needs met.

Fortunately I can **RECOGNIZE** almost the instant my Character 2 has possessed my body and consciousness. Character 2 feels heavy, burdened, or desperate, just as though doomsday has perilously arrived. Character 2 feels prickly in my body and tense in my throat, and the left half of my jaw aches with pain. I become highly focused as my head feels like it is in a deep cloud, and I feel sequestered from my other characters. Shame, guilt, embarrassment, and the impetus to blame someone else for something that has happened are old patterned responses that my Character 2 has perfected in her attempts to fend off threats. It is critical that we each train ourselves to immediately **RECOGNIZE** when our Character 2 has come online so we can take full responsibility for managing and disarming its pain and potential wrath before we express them in the world and sabotage our relationships or lives.

I de-escalate the hostility of my Character 2 by **RECOGNIZING** that it has been triggered, and then I immediately push the pause button. I purposefully calm this character down before I say or do things that I might later regret. Feeling shame or guilt about my

Character 2's wiring only vexes my attempts to heal her, so when she is triggered, I practice the 90 Second Rule. It works like a time-out, or a count-to-ten, giving me a moment to reset. Under extreme duress my Character 2 **RECOGNIZES** she has been hijacked, and she brings the word B-R-A-I-N into her mind to help her remember she is not alone.

In contrast, I **RECOGNIZE** that I am in my ebullient Character 3 when I am feeling really animated, a bit jittery, or just bouncing out of my skin with exuberance. Nothing floods through me like a flush of adrenaline, and I can instantly **RECOGNIZE** this explosion of energy. When I feel present, attentive to the moment, and wanting to play or connect with others, I **RECOGNIZE** that I am in my Character 3. Also, when I am feeling artistically or musically creative or curious, my Character 3 is center stage.

But Character 3 is not always happy, in that sometimes the emotional surge of an immediate danger can trigger a right-brain alert response that is easily **RECOGNIZABLE** as well as easily distinguished from my Character 2 threat response. While my Character 2 feels heavy, drained, or distraught when emotionally triggered, Character 3 feels like an energy surge of blood rushing instantaneously into my limbs, preparing me for a fight-or-flight response. Feeling like I want to jump out of my skin, I have an odd sense of acceleration, as a flood of prickly heat flushes up my spine and explodes like a volcanic river as a purge of sweat.

Although this triggering of my Character 3 can be startling and uncomfortable for anyone nearby, the pain I feel internally is equally alarming to me. Because I understand that this reactivity is wired in my brain to save my life, once I feel the floodgate open and I **RECOGNIZE** it for what it is, I do my best to step back and let it dissipate in private. Our prisons are filled with first-time offenders whose Character 3 had a momentary full-blown aggressive attack. Imagine how having had the ability to call their Four Characters into a Brain Huddle could have saved them from acting out and doing something they would later regret.

Character 4, in all of its glory, is always a delight and easy for me to **RECOGNIZE** and enjoy. When I am feeling total contentment, an open expansiveness in my chest, and a deep appreciation for whatever life has brought my way, it is simple for me to **RECOGNIZE** that my Character 4 is the circuitry that I am running. In this mindset, my heart feels calm and I am filled with a deep sense of gratitude for everything just as it is, even if another part of me might wish something were different. I can **RECOGNIZE** that Character 4 is always present and available for me to tune in to, even if I can't sense her when my Character 2 has commandeered my focus.

When I **RECOGNIZE** the character that I am being, no matter which one it is, I validate that character and my link to its awareness becomes stronger. When I pay attention and care enough to know which character I am embodying, I connect with it, and simply by doing so I connect with myself. When I **RECOGNIZE** the value of my Four Characters, I don't need anyone outside of me to validate me. Instead, I am clear about who my Four Characters are, and I know that I am enough. I know that I am not only worthy of being loved but that I am love.

The same is true for how I connect with you. Only when I care enough to pay attention to **RECOGNIZE** which of the Four Characters you are inhabiting at any moment can I authentically connect with you. If I want to fully see you and validate you, I must first **RECOGNIZE** which character you have brought into our encounter. If you come in as a Character 1, you would probably rather receive a *you did a good job* affirmation than the *you light up the world* type of praise I might give your Character 4.

And if you come in as your Character 2, perhaps especially if you come in as your Character 2, I need to **RECOGNIZE** that, accept that you are in pain, and shift my affect to match your energy, welcome you, support your need to be heard, and love on you. I need to **RECOGNIZE** if your pain is hurt or sadness so I can show up as my Character 4 to soothe you. Or if your pain is fear or anxiety, maybe my Character 1 needs to come out to protect you. If you come in as your Character 2 and I dismiss your need, I miss a golden opportunity to sincerely connect with one of the most tender and vulnerable parts of who you are. Our willingness to **RECOGNIZE** which characters are in the room is our pathway to genuine connection and in-to-me-see (intimacy).

A = APPRECIATE

APPRECIATE whichever character is present and **APPRECIATE** the reality that all Four Characters are always around, whether we are aware of their presence or not.

When I **APPRECIATE** the innate value of whichever character I am exhibiting in any moment, beyond merely recognizing that character, I am intentionally honoring and respecting the skill set of that part of me. When we focus on, validate, and **APPRECIATE** the strengths of each of our Four Characters, we are empowering our ability to engage with them. Nothing is more important than our respect for ourselves, or more connecting with others than our respect and **APPRECIATION** for the gifts that each of their Four Characters bring to the table.

In addition, during my worst bouts of anxiety, fear, or anger, when the alarm circuit of my Character 2 is triggered and I feel cut off from my other characters, just being able to **APPRECIATE** the fact that they are there reassures me that I am fine. Even in those moments of despair and hopelessness, I know my other characters will come back online as soon as the energy dissipates back into those parts of my brain.

When I **APPRECIATE** that my Character 2 can be a bit awkward (loud, attacking, inappropriate) in how she sounds my alarm, and I remember that she behaves in this way because she loves me and is trying to protect me and does not know a better way, it makes it a lot easier for me to honor how courageous she is in the face of those perceived threats. Keeping close tabs on her and **APPRECIATING** her intention and efforts help her feel safe.

When I move into the Brain Huddle and **APPRECIATE** each of my characters individually, I **APPRECIATE** that my Character 1 wants to be the authority and take control of the details of my world so I can live my best life and know that she has me under her wing. She is excellent at what she does, and I am grateful that she is a master of many things, including the management of space, events, and people and their schedules.

My Character 2 is a faithful servant that shows up to protect me, and I **APPRECIATE** her willingness to play that role so I can be safe and flourish. When I **APPRECIATE** whatever emotion I am experiencing, it brings richness to my life, and I am grateful that my Character 2 has been willing to step out of the consciousness of the present moment so I can experience the linearity of a past, present, and future. When I **APPRECIATE** and value myself, I operate from a position of power. When I deny my emotions, I fuel my discontent and internal struggle.

Character 3 embraces the thrill of my existence, and I **APPRE-CIATE** that I have the ability to savor every fleeting moment and experience. On top of that, I so value and **APPRECIATE** her open heart, and her eagerness to play and genuinely connect with you.

My Character 4 rests without judgment in the grace of all that is. And I **APPRECIATE** the 50 trillion molecular geniuses that work seamlessly together so that not only do I have the ability to exist at all but you are here for me to share with.

And I **APPRECIATE** all of your characters for the same reasons I **APPRECIATE** all four of mine. It is our **APPRECIATION** for one another that seals our connection. Like plugging our phone into the charger, we can recognize one another, but if we don't line up those little teeth to make a solid connection, the energy will not flow from the wall socket into the phone. You may recognize what character I am displaying, but when you sincerely **APPRECIATE** its value and offer gratitude for its existence, that is when we align with one another and manifest a lasting and meaningful connection.

I = INQUIRE

INQUIRE within and invite all Four Characters into the huddle.

Once I have pushed the pause button and become aware of my BREATH, and then RECOGNIZED which of my Four Characters I have been embodying, as well as RECOGNIZED which characters of others are around, I have then offered APPRECIATION up for all of our characters. At this point in the Brain Huddle, it is time for us as a collective of all Four Characters to **INQUIRE** about what our best next move might be. We **INQUIRE** when we are curious, and we are curious when we care. Using the tool of the Brain Huddle, our Four Characters come together to voice their opinions. We **INQUIRE** by first observing ourselves, second by observing those around us, third by observing ourselves in response to those around us, and fourth by observing how others are responding to us.

Say, for example, I walk into a room where a couple has been arguing as two Character 2s. In that moment it would be appropriate for my Four Characters to privately take a Brain Huddle to INQUIRE about what we might consciously choose to do next. Since my Character 4 has already ascertained the tension in the air, it might be good for my Character 3 to hop in with some comic relief. However, if I don't know these folks very well, that strategy might fall flat. Choosing to bring my Character 1 on board instead, so she can address the logistics and perhaps offer some assistance if needed, might prove to be the best plan.

Not long ago, I was driving down the highway when the car in front of me swerved and accidently hit a rabbit. As I approached I saw the animal was clearly hurt but had not yet crossed the rainbow bridge. In that moment I felt myself flood with an array of intense emotions, and I instinctively invited my Four Characters into a Brain Huddle. During the INQUIRY, they all had something to share. Little Abby had been instantly flooded with enormous grief and reactively didn't know what to do other than freak out. Character 1, Helen, tried to determine if she should turn around to try to help the animal or just keep going forward because there was an impatient driver tailgating us. Pigpen wondered, *Huh, I don't have a rabbit in my mammalian brain collection*, while Queen Toad closed the INQUIRY by choosing to hold the rabbit in a tenderhearted place from afar. All four of my characters agreed to sit in quiet prayer, sending love to the creature whatever its fate was.

If I am showing up as a Character 1 most of the time, it is not my nature to choose to INQUIRE very often. Since our Character 1 is busy getting the job done rather than exploring new possibilities, encouraging our Character 1 to pause, shift into huddle mode, and then INQUIRE with the others about what their opinions are is generally a helpful idea. When I do INQUIRE, I am curious about how I am showing up, and I am curious about who you are showing up to be as well. When we INQUIRE and are being curious, we are engaging our right-brain Characters 3 and 4, and these two characters can provide a refreshing perspective and neurological reset.

INQUIRY is a great gift that we offer others, as it lets them know we are available for a real connection. When we INQUIRE into both ourselves and others, we accept, encourage, and invite everyone's Four Characters to participate in both the huddle and in choosing the next strategic move in the game of life. This is particularly important when someone is in discomfort or feeling the need to be right, as it invites everyone's Four Characters to participate with their unique insights. By taking responsibility for the character I embody, I take responsibility for the energy I bring into a room.

N = NAVIGATE

NAVIGATE our new reality with all Four Characters bringing their best game.

For the Brain Huddle we began by consciously choosing to pause and take a deep BREATH so we could bring ourselves to the forefront of our mind in the present moment. Then we RECOGNIZED which character we were exhibiting and expressed APPRECIATION for that specific skill set, as well as appreciation for the awareness that all the other characters were available for action. Next, all four of our characters joined together in the Brain Huddle for an intensive INQUIRY about current circumstances, and now it is time for our Four Characters to NAVIGATE our new reality as a team.

Life is a moving target, with change as the only constant. The Brain Huddle allows us to consciously shift away from our natural tendency to react on automatic and purposely take responsibility for who and how we want to be. Our circumstances are constantly changing, so choosing a static response to a moving target will inevitably result in failure.

If we are to be successful, our Four Characters need to be available to NAVIGATE our life and be flexible with whatever characters show up in others. Say, for example, I just bought a shirt that had a stain on it that I did not realize was there until I stepped out of the store and into the parking lot. In that moment all four of my characters had an opinion, of course. Abby felt disgruntled and put out,

Helen needed to go right back into the store and replace it immediately, Pigpen was happy to look at all the sparkly stuff some more, and Queen Toad knew we had plenty of time so all was well.

During INQUIRY, my Four Characters decided that this was a perfect task for Helen, so back into the store we went with Helen in the lead to exchange the item for an identical one. At the customer support desk, we could have encountered any of the Four Characters. A Character 1 clerk might choose to escort me back to the shirts and help me find a replacement, because they value me as a customer and want me happy. A Character 2 might be a stickler for the rules and insist that I fill out the paperwork for a return. Then I would have to go find a new shirt and wait in line again to buy it. A Character 3 or 4 customer service clerk would probably just wave me through so I could replace the item without any hassle. Each of my Four Characters would **NAVIGATE**, with flexibility, whatever character we encountered. If it turned out that we were working with a Character 2, Helen would purposely keep Abby at bay and breathe deeply, while Queen Toad might choose to say something complimentary and kind to the person. We **NAVIGATE** who we are moment by moment, depending on who we encounter.

We really do have the power to choose who and how we want to be in the world, and we have so much more power over what is going on inside of our brains than we have ever been taught. In his wonderful book *Getting to Yes with Yourself,* my good friend Bill Ury shares the strategy of "going to the balcony" when it is time to undertake, or **NAVIGATE**, an intense negotiation. In the language of the Four Characters, Bill is encouraging people to step into the consciousness of their Character 4 so they can bring their big-picture perspective into the conversation. When we are willing to explore what both sides have in common, and find our way past our emotion, we can **NAVIGATE** our way into an outcome that feels like success for both parties.

The Reset

As we have explored, the Brain Huddle can be a fantastic tool for our Four Characters to use on a regular basis when we are not in distress. I use this tool many times a day just to get a clue about how I am showing up in my life. By strengthening this circuitry and using this tool regularly, I know that I can call on any of my Four Characters at any time. And I have to admit, it's fun watching my characters in the wild.

In addition, I have shared how I have used the Brain Huddle tool during extreme duress, and how it has provided me with a lifeline when I needed it the most. The B-R-A-I-N acronym itself reminds me that all of my characters are always available, even when I can't feel them. It anchors my knowing that I am not alone, and that I am okay.

The Brain Huddle is also a powerful tool for resetting our connection with another person in a moment of conflict or distress. It is important to remember that inside every relationship or encounter there are eight of us vying for the microphone. When all eight of our characters are trying to get along, it is not unusual for one of our Character 2s to be triggered unexpectedly. And once our emotional Character 2 has been triggered, although it may take only 90 seconds for that emotion to flush through and pass out of us, if we reengage too soon, when we are still feeling vulnerable, we may end up being retriggered.

When we fight with another person, it is important to recognize that the energy in the space between us has become just as charged as we are. Turning off neuronal circuits is somewhat similar to turning off electrical circuits, in that it takes a little bit of time for the energy to completely dissipate and neutralize. It is important that our Four Characters take some time to completely reset before entering back into a tense or toxic setting.

Thus, creating physical space between us and the other person is generally a really good idea, even if it just means stepping into another room. Next, if both parties are aware of their Four Characters, and each is willing to take a Brain Huddle, this is a good first step to renewing a positive connection. If even only one person is willing to step out of their Character 2 and engage in a Brain Huddle, there is hope for a reconnection. However, I will repeat: two Character 2s will never find resolution until one of them is willing to step out of that character and embody one of their other characters.

Whether you find yourself challenged in an encounter with another person or emotionally triggered and navigating a stressful situation on your own, the Brain Huddle offers a powerful opportunity for a reset. Besides proving to be a fantastic tool for normal healthy living, the B-R-A-I-N acronym has the power to beam like a neon light through the blinding fog of a devastating bout of negative emotion. If you ever need to rescue yourself from a desperate situation, let this be an anchor.

COMING UP NEXT: THE FOUR CHARACTERS IN THE WILD

Now that we have examined each of the Four Characters in detail, as well as the benefits of bringing them together in the Brain Huddle as a tool for both crisis management and daily living, let's take a closer look at the Four Characters in different areas of real life. I like to call this "the Four Characters in the wild."

In Chapter 9 we will delve into the intimate relationship between our Four Characters and how they relate to and take care of our body, in both wellness and in illness. Unquestionably our most important relationship is the one between our brain and body, and all of the Four Characters are quite predictable in how they look at, attend to, and nurture this vital relationship.

In Chapter 10 we will consider the Four Characters and how they interact in romantic relationship with the Four Characters of a potential partner. Our whole world revolves around how we connect with others and how they connect with us. The Four Characters have different values, and they live our lives and establish our relationships with one another based upon those values. In this chapter we will take a peek at who might be attracted to whom, which combinations are likely to work, and some of the dynamics we can predict based on what we know about the Four Characters. You will probably recognize some of your own patterns, and hopefully gain a little insight into your own relationships.

Chapter 11 views the Four Characters from a completely different perspective. It is one thing for us to approach this material through the filter of a healthy brain, but in this chapter we will talk about what is going on in the brain that is managing alcoholism, addiction, or recovery. The Four Characters material up to this point has been all about helping us create healthy relationships within our own brain, as well as between our brain and others, through the creation of healthy connections. Because drugs and alcohol cause brain cells to disconnect rather than connect with one another, they not only interfere with the operation and function of our own brain, but they also create disconnection in our interpersonal relationships and have the power to derail them.

In Chapter 11 we will explore how understanding the Four Characters might help a person create the conditions necessary to successfully navigate a recovery program such as the Twelve Step Program of Alcoholics Anonymous (AA). Thanks to our brain's neuroplasticity and neurogenesis, we have the ability to heal and recover from all sorts of trauma, including drug and alcohol abuse. As we dig more deeply into what is going on in the brain for a successful recovery at the level of the Four Characters, we will discuss how the AA Twelve Step Program, the Brain Huddle, and the Hero's Journey are in many ways the same journey toward healing.

Finally, in Chapter 12 we will expand our vision of the brain and our Four Characters into the bigger picture of humanity. We will take a look at the evolution of our Four Characters over the last hundred years in the U.S. through the filter of our generational differences. In addition, we will look closely at the influence technology has had on both our brain and the expression of our Four Characters. We will walk away with a better understanding of how we are neurologically different from our children, especially at the level of our Four Characters.

As you read through the following chapters, regardless of the specific subject, I think you will find the behavior of each of the Four Characters to be both consistent and predictable. In addition, you will probably recognize your own attitudes and behaviors quite easily. It is important to note that you may readily identify with multiple of your characters under different circumstances. Although we each may have a dominant character, we really do tend to exhibit all Four Characters in different situations. Considering there is no right or wrong, I am hoping what you learn about yourself will be fun for you.

Please remember that although you may recognize some of your tendencies from past behavior, you are not bound to your old programming. Many of us have been running on automatic for such a long time that our health and well-being may have suffered. Now that you know you have the power to choose, and you know what your Four Characters options are and how to bring them together in a Brain Huddle, you might choose to make different decisions. For example, if you have traditionally managed an illness with your Character 2, you may now choose to let your Character 1 take over that role. Or if you have young children and tend to parent with your Character 1, you might try shaking things up a bit and show up doing a Character 1 task as a fun-loving Character 3 instead.

As I mentioned early on, we have more power over what is going on inside of our brain than we were ever taught. History is filled with examples of remarkable people who have not only endured and survived horrible treatment and events but have come out of it emotionally and cognitively empowered. Following a beating for civil disobedience, Mahatma Gandhi is said to have professed, "Nobody can hurt me without my permission." Essentially, he was declaring the realm of his brain as his own domain. This is the ultimate power of our Four Characters.

Part III

THE FOUR CHARACTERS IN THE WILD

CHAPTER 9

CONNECTION TO OURSELF—OUR FOUR CHARACTERS AND THE BODY

Without question, the most important relationship we have is the one between our brain and our body. How do the Four Characters see this vital relationship?

- Character 1 sees our body as a vehicle.
- Character 2 sees our body as a responsibility.
- Character 3 sees our body as a toy.
- Character 4 sees our body as a temple of the soul.

CHARACTER I

Character 1s see their body as a vehicle that they use to accomplish things in the world, so they keep a close eye on how well their machine is running. Comparable to how they tend to their cars and other mechanical possessions, Character 1s are committed to getting a physical every year to make sure everything is functioning properly. Character 1s like information and believe knowledge is power. Their goal is to nip a problem in the bud, so routine checkups are purposeful.

Character 1s care about building relationships with their doctors because when there is a problem, they are interested in consulting with an authority and then becoming an expert. Character 1s don't mind spending money for a job well done, and they are curious about their health and interested in maintaining it. Because they pay close attention to the performance of the machine, Character 1s have pretty good body self-awareness. They tend to notice how they feel, and if something does not feel right, they have it checked out. Character 1s take personal responsibility for their body and do not turn that responsibility over to their doctor, coach, or personal trainer.

Character 1s participate in wellness screenings and are good at scheduling appointments up to a year in advance. They are faithful at taking supplements and will get up and go to the gym and lift those weights that they hate to lift if they think doing so will help. Character 1s build that routine into their schedule because they take pride in what is right about their body. At the same time, because they are perfectionists and tend to be critical of their appearance, they are highly motivated to stay fit and look good. This group plans on living a long, quality life, so they take it upon themselves to do what they believe they need to do to take care of themselves.

Character 2

If our Character 2 is in charge of our relationship with our body, we will manage our care in a way that is the complete opposite of our Character 1. Because our Character 2 has very little body awareness, everything health related feels scary and unknown. Character 2s see knowledge of the body through the lens of doom and gloom because a million things can and will go wrong. To the Character 2, the whole medical world is only bad news, never good news, and every little concern is seen as a potential death sentence. Since they find the concept of death terrifying, they may die a thousand deaths inside their minds, long before it's actually their time to go.

As a result, when our Character 2 comes online in a situation that involves our health, we strategize our body care in one of two ways. We either stick our head in the sand, dread the annual examination, and resist going to the doctor, or we make mountains out of every molehill and become a frequent visitor to the emergency room. To compound the problem, our Character 2 will gather and share woeful stories with others about how someone back home had that problem, and then their arm fell off. Character 2s live in such a high state of stress and worry that when they are not in control, their bodies respond with accelerated wear and tear.

Character 2s are not interested in attending wellness screenings or joining a gym, but they often complain about their aches and pains, without prodding, to just about anyone who will listen. Because our Character 2s don't like to tend to body complications in a timely fashion, they trot off to the ER for chronic pains but afterward do only the minimum that they have to do to help themselves. If they are motivated by a friend or a work wellness program, they may walk around the building for 10 minutes, but as soon as they feel uncomfortable, they will stop trying. For Character 2s wellness is all about what is going wrong rather than what is going right. As a result, the Character 2 will sit on the sidelines and be the first to tell you all the reasons why it is not a good idea for you to be doing something active.

Just as the Character 1 wants to meet directly with the physician because they want to become an expert about their condition, the Character 2 is happy to share with the more readily available nurse practitioner. The Character 2 values having a medical audience to whom they can share their woes more than they care about the medical professionals' credentials.

Character 3

Character 3s get excited about anything medical because it's interesting and cool. A Character 3 says, "Wow, look at my vitals!" The body is a toy to a Character 3. It is their playhouse, so they want to use it, challenge it, and take good care of it. The body is such a curious thing to a Character 3: "I mean, just look at my toes! When I am excited, they waggle back and forth just like a tail! Look how high I can jump! Look how fast I can swim!"

More than any of the others, Character 3s have a deep and intimate body consciousness. They care about strength training, quality of performance, and perfecting their timing. Character 3s are curious about how much they can achieve with their body and how well they can get it to perform. Character 3s live in their body and will push it toward its optimal performance. "I know I can hike that trail in eighty minutes, so can I hike it with twenty pounds on my back in ninety minutes?" Fitness is fun and a great way to pass the time for a Character 3.

Although Character 3s may not have an annual exam on their calendar, they will take advantage of local community wellness programs as they happen upon them. Character 3s celebrate their body, so living a highly active lifestyle is natural for them. For exercise, instead of just hitting the weights in a gym, Character 3s will go out and do something fun and adventurous. They would rather build a stone path where every step weighs 90 pounds, or master a climbing wall in the local park, than work a routine in the gym. Overall, Character 3s have a higher rate of emergency visits due to acute breaks and misfortunes because they are out pushing their bodies on a regular basis, and accidents happen.

Character 4

Character 4s see the body as the holy temple of the soul. As such, they are grateful for the miraculous gift of life, and they function inside a wellness frame of mind whereby they accept the responsibility of tending to the well-being of the temple. Character

4s nurture their mind/body/spirit by engaging in self-care. They embrace holistic and alternative methods of nurturing their body and stimulating their senses.

Massage, yoga, essential oils, and other holistic practices speak to the heart of the Character 4. When possible, they are members of food co-ops and eat organic foods in an attempt to minimize their chemical toxicity. Character 4s are active members of their community, and it is important to them to support their local farmers' market. Gluten and animal products may or may not be a part of their menu, and they are open to taking natural supplements.

Character 4s believe in going to their local acupuncturist, osteopath, chiropractor, or neuro-movement specialist as a part of their wellness routine, particularly when something needs attention. Character 4s are committed to going outside for a walk on a beautiful day or purposefully stretching on the floor in front of their TV. You will find Character 4s hoe-dee-doing around the park or neighborhood with a friend or pet. They will pause and enjoy a meaningful chat with a squirrel, hug a familiar tree, and open their hearts to the nature around them, as they are deeply connected to all life. Character 4s will not just remember to bring goodies to feed the wildlife at the park, they will stop on the way to buy extra. These random acts of kindness are important to the Character 4 and directly contribute to their overall well-being.

How the Four Characters Manage an Illness

Character 1

When it comes to managing an illness, Character 1s are informed patients. Their linear thinking kicks in, their rational minds turn on, and they study to become a specialist. The diagnosis becomes their full-time job, so they assess the problem quickly, learn everything there is to know about it, and then manage the situation with absolute precision. Take the example of a chronic illness such as type 1 diabetes. To manage this disease, Character 1s will change their diet and avoid eating sugar at all costs. They tune in to their body and want the most recent technology and analytics, so they will invest in the most accurate continuous glucose monitors and insulin tubeless pumps that will deliver instantaneous analytics through a phone app.

Character 2

Our Character 2 already sees the medical world as all bad news, so facing a real illness like type 1 diabetes is enough to overwhelm them and render them paralyzed with fear. As a result, they succumb to the feeling of being distraught. They stick their head in the sand and try to ignore the problem for as long as they possibly can. These are the folks who will sneak sugar and never fully embrace a healthy management of their own protocols. It's not that they don't care, but to the immature Character 2, fear and anxiety can be so overpowering that these people cannot think clearly about what they need to do to help restore their health.

Some very healthy Character 1s, 3s, and 4s may find that they turn into a petrified Character 2 when they become ill, simply because of the fear many of us harbor about dying. When we do devolve into our Character 2 around a severe illness, we become more interested in what we can get away with, rather than what we might do to help ourselves. Remember: our Character 2 represents our potentially self-destructive five-year-old.

If we do dive into our Character 2 out of fear of a diagnosis, hopefully there is a responsible Character 1 around who is willing to help us, or a Character 4 who is available to nurture us. Oftentimes, however, our Character 2 may try to force the impossible job of managing our illness onto someone else. But of course, they would get all of that responsibility without an ounce of our cooperation. To make things worse, our Character 2s are not interested in messing around with the latest medical technology. Partly because we would feel tethered to the machinery, and let's face it, we can't cheat on our sugar intake if someone is monitoring the data.

**Character 3s get excited about anything
medical because it's interesting and cool.
A Character 3 says, "Wow, look at my vitals!"**

Character 3

Character 3s will minimize the severity of the diagnosis and say, "It's not that big of a deal." Because they don't want to give up their sweets, Character 3s will find a way to work around the problem and find sugarless candies and cookies that won't spike their sugar. They will explore the influence of substitute sweeteners on their sugar levels and will want to use the latest tubeless insulin pump technology because it's cool, fast, and easy. Besides that, Character 3s are not disciplined enough to test their blood sugar every two hours, so they view the latest technology and phone app as a ticket to free living that they can take with them everywhere they go.

Character 4

Character 4s want to know what their options are and will make friends with their diagnosis. This character is all about the health and wholeness of their body/mind/spirit. Character 4s will take responsibility and use alternative medical options to manage their problems. They will visit naturopaths, chiropractors, acupuncturists, and various types of energy healers. For diabetes type 1, like Character 3s, Character 4s will explore the influence of honey, agave, chicory, and coconut sugar on their sugar levels. They want to be easy on their body, so they will use meditation to lower their stress, blood pressure, and blood sugar levels.

Character 4s will set a routine for regular movement and exercise to improve their medical numbers. They will welcome the ease that new technology offers and explore the latest in testing and pumps. Character 4s show up for themselves and accept what reality is, and they will do what they need to do to manifest a positive prognosis. It is the Character 4 in each of us that will embrace with gratitude the silver lining of having had the chance to have this challenging experience.

How the Four Characters Manage Fitness, Diet, and Dieting

Character 1

When it comes to taking care of their body, Character 1s take responsibility and hold themselves accountable for the shape they are in. When they hit their top weight, they take action to not go one ounce over the limit. Character 1s are disciplined people, so when it comes to exercise and portion control, they do what they need to do to keep their vehicle running well. They are busy people and have a lot to accomplish, so it is natural for them to address their body consciously.

When they want to lose weight, Character 1s are going to count food points, be disciplined, and diet effectively. They will set up their kitchen with everything they need and learn the system and be good at it. They will use cleanses to jump-start their diet and suggest that their partner, or others in the house, join them on the program, since they are good at helping and supporting others.

Character 2

Our Character 2s become intimidated by the number on the scale and look for the quick fix. Any diet is seen as pure punishment, because it means that we must deprive ourselves of something we want. As a result, Character 2s buck the system, cheat, and do not succeed because they see themselves as powerless over their desires. Character 2s will exert the minimal amount of energy, and as soon as they become uncomfortable or don't see success, they will bemoan the process and indulge in more calories. If a Character 2 does lose any weight, they will complain, "Oh gosh, it almost killed me!"

Character 2s are not disciplined enough to take on a complex point-counting system, but they will try the latest fad pills or electronic muscle stimulators. The Character 2 is willing to order prepackaged food because they don't want to exert any real effort. Character 2s will starve themselves and then binge and then starve themselves again. If they are lucky, they are living with a Character 1 who is willing to do all the work for the point-counting system, and under these conditions they may actually succeed, but only if the Character 1 polices the pantry.

Character 3

Character 3s will figure out how many donuts or how much ice cream they can eat and then exercise enough to ensure they do not gain a pound. This is the character who is going to binge like a maniac in the present moment, ruining the last three days of their diet, but then hike five miles a day for the next three days to make up for it. Character 3s will eat the entire bag of potato chips and then eat nothing except vegetables for the next few days. If motivated to drop some pounds, they will eat smoothies, protein shakes, or healthy granola bars.

Character 3s have a good sense of how their body burns the energy they put into it, so although they will not be motivated to count calories or points, they will limit what kinds of foods they eat based upon how those foods make them feel. Character 3s are very sensitive to the influence of what they eat and how that impacts their energy level. Character 3s will do the South Beach Diet simply because it sounds like fun: "Yay, I get to go to the Beach! Hey! I'm doing the Beach, what about you?" They will eat prepackaged food because of the convenience but then binge by having a veggie burger with no bun, just to see how far they can go on the diet without being totally compliant. Character 3s will eat everything in sight, in general, because they consume, consume, consume and are always hungry because they are always in motion.

Character 4

Character 4s are looking for a healthy balance between work, family, play, friends, and time with their Higher Power. They go to yoga class, attend retreats, get massages, and meditate. They generally don't want to consume anything that has a face, or breathes, so they are often vegetarian, vegan, or they at least stick to an organic diet. To balance a minor weight gain, they will cut back on calories through portion control or eat more fruits and vegetables.

Character 4s know that you can never out-exercise a bad diet. They feel better when eating more protein or high fiber and will change the entire way they eat if they need to. They consume a whole-foods diet, and everything is in moderation. Character 4s conceptually carry that donut all the way out, from the yummy taste bud to seeing that donut right there on their left hip.

How the Four Characters Relate to Medical Professionals

Character 1

Character 1s are interested in following the most direct path to a solution, so they will beeline themselves straight into their personal doctor's office. Or they will make a phone call to a medical line or use a tele-doctor. The Character 1 wants the M.D. degree with the biggest solution for the problem. They want to speak with the expert because they want to learn everything they can about the problem and become an expert on the issue themselves.

When it comes to compliance with a course of treatment, Character 1s are equally direct. They will follow the doctor's orders and work the designated program. They will go above and beyond the minimum effort to get well. They will change their exercise routine as well as food choices with minimal complaining. Character 1s see the need for change as a challenge that is worth the outcome of regaining their health.

Character 2

Our Character 2 will use the emergency room because they are dying every time they feel bad, and they will bring the 20-pound solution to the 5-ounce problem. Or they will be on a first-name basis with the clinic's nurse practitioner, whom they will seek out often. Character 2s want all the tests they can get just to make sure nothing is missed, but because they look for the worst possible outcome, they become overwhelmed.

Character 2s will Google the morbidity rate of a hangnail and then bring it up as casual conversation at the dinner table to anyone who will listen. Although Character 2s want the confirmation that they are really sick, they don't want to hear the protocol to regain full health, because the whole medical/body phenomenon is terrifying and intimidating. Anything the doctors have to say about life changes will simply limit their quality of life, so they don't really want to hear it . . . but yet they do . . . but yet they don't.

Character 2s see dietary and lifestyle changes as a punishment, a loss of freedom, a limitation to their quality of life, a sacrifice, a pain in the neck, and just one more thing they have to manage. If they even bother to try to stay on a medical protocol, they will not be enthusiastic about it or do it well.

Character 3

Character 3s do well with the convenience of the walk-in care center. They want the "Doc in the box" so they don't have to schedule an appointment but can slip in on a whim. These urgent care centers get you in and out quickly, and the Character 3 likes that. They tell you what you need to know in order for you to get better, and you don't have to establish a relationship with the doctor. If you meet with a nurse practitioner, that's okay too, as long as they have the expertise to meet your need.

Once they've received a prescription or a course of treatment, Character 3s will use modern technology to set a schedule, and they will try to find a way to work around the problem. They may be strict about some things and push the envelope with others. Character 3s will set an alarm on their phone to remind them to take their meds on time, and then figure out what is the outer limit of what they need to do to be successful. Like Character 1s, Character 3s are looking for best practices and will find a way to win. Character 3s may join a support group because they like to collaborate with others, as they thrive on conquering a situation through a team effort.

Character 4

The holistic approach appeals to the Character 4. They think in the big picture of the whole body and want to know the minimum they need to do to get the maximum return for their efforts. The Character 4's focus is on prevention, and they will seek alternative types of practitioners to promote their own healing. Character 4s are good with less invasive alternative providers and tend to manage their wellness long before any illness sets in.

Character 4s will get on board with the medical establishment just enough to pacify the acute situation and then explore alternative options to manage a long-term problem. For example, instead of taking a statin to lower a high-cholesterol issue, the Character 4 will master the combination of honey and cinnamon. Instead of sticking to an intense physical therapy program that would rehabilitate their rotator cuff to its limitation, a Character 4 would work with an Anat Baniel NeuroMovement practitioner over the long term to regain complete function.

How the Four Characters Age Well: Physically, Mentally, Emotionally, and Spiritually

Character 1

Character 1s take care of themselves and do what they need to do throughout their lifetime to maintain the machine. They are conscious of their body and hyperaware of their aging. They notice the need for maintenance and exercise because they want the result, not for the love of the game. Character 1s will take care of themselves and nip and tuck as they age, and they would be the type to keep a pair of shorts from their college years, just to see if they can still fit in them.

If a Character 1 splurges during the holidays, because they have rigid boundaries, they will work hard in the New Year to lose what they gained. If a Character 1 has a joint replacement, they will educate themselves on the best doctors and the latest in prosthesis. They will take responsibility for their physical therapy, practice between sessions, and enthusiastically adopt the new device as a better part of themselves. They will then get back into life and continue their exercise routine.

As they get older, many Hard Character 1s will pivot from expressing harsh judgment to expressing a softer sense of gratitude. With this shift comes a natural deepening of their relationship with their body, with themselves, with others, and with their Higher Power. Life tends to whittle away at our edges, and with retirement come new opportunities and choices. Living life as a Hard Character 1 is isolating, no matter the age, because they have not nurtured a loving tribe or community. In the long run, the financial wealth a Hard Character 1 may have acquired can buy a lot of things, but you cannot buy a loyal fishing buddy or kindness from a palliative-care nurse.

Character 2

Because the cells of our Character 2 are programmed as our fight/flight/freeze sympathetic nervous system, we are neurologically wired to bring information in from the present moment and then compare that with every experience we have ever had. Our Character 2 cells sort and match the worst possible outcomes and then inject that insight into our present-moment consciousness. As a result, we fall victim to our feelings of loss, pain, fear, anxiety, and threat, simply because this part of our brain is wired to do so.

To complicate matters the cells making up our Character 2 are the same cells that are specifically designed to perceive ourselves as a single solid entity, separate from everything else around us. Long gone are the days when we symbiotically existed in our mother's womb, and gone are the days when all we could perceive was the enveloping feeling of cosmic love.

When it comes to our health and well-being over the expanse of a lifetime, if our Character 2 is in charge, wellness will be hard-earned, if possible at all. The fear that our Character 2 brings to the forefront of our consciousness can be strangling and constricting. Without any discipline or set upper limits on weight gain, no one takes ownership or responsibility for the body. With increased age older Character 2s will live in the glory days of times past and share stories about how they used to be able to do this and that.

Character 2s feel their aches and pains and then use them as an excuse to not do anything more. They focus on what they cannot do rather than what they still can do. When they have surgery to get a new body part, like a new joint, Character 2s are afraid to use it. Although they will go to physical therapy, they will put in the minimal amount of effort they can get away with and then have limited recovery and blame their failure on a bad system.

If a Character 2 is going to age at all gracefully, they must have a change of heart and mind and be willing to let their other characters step in to create some ease in the matter of their health. Personally, I hate going to the doctor now because one day way back when (worst-case scenario), I had to have my head cut open, and who on earth would ever want to see something like that happen again? My Character 2 believes that if I don't go to a doctor, no one will find anything wrong with me and I'll never have to go through something like that again. Right? Wrong. My Character 2's fear is so powerful that it gives me a million excuses to not abide any preventative care.

Clearly the price we pay to live with the belief that we are individually independent people who have free will is the consternation of our Character 2. Fortunately that part of our brain coexists with our Characters 1, 3, and 4, who all have the capacity to interact with, intercept, and integrate our behaviors. None of us is getting out of this life alive, so once all of our Four Characters agree that our ultimate goal is to live well, and die well, we can all weigh in on what we need to do to make that happen.

Character 3

Character 3s can run wild and crazy with our health during our youth, and hopefully get away with it, but somewhere along the line we need to recognize our limitations. As youngsters we didn't measure how much water we were drinking, we just drank to satisfy our thirst. As we age we need to place a little more focus on health rather than fun and calculate longevity versus risk. As our Character 3s become more aware and intentional about what our bodies need, and what limitations we need to set on our risk-taking behaviors, our overall health will thrive.

Character 4

As our Character 4 ages, we need to pay closer attention to what our body is telling us and then engage in behaviors that will increase our brain's awareness of our body. The more powerful our divine Character 4 becomes, the less well connected we are to the physical world. Continuing a practice of yoga, tai chi, or anything else that will help us ground ourselves in our muscle groups will help us stay connected to the physical world. As we age, our minds naturally shift more into the realm of the mysterious, so intentionally choosing to increase our brain's discernment of our body is a really good idea.

Now that we have examined how our Four Characters relate to our body in both wellness and illness, let's expand our perception and take a closer look at how the Four Characters predictably respond to, relate to, and interact in relationship with others.

CHAPTER 10

Connection with Others—Our Four Characters in Romantic Relationships

We humans are social creatures, yet for many of us, creating healthy relationships is probably our greatest challenge. No matter how tolerant we may be of our own unique behaviors, living with someone else's oddities adds a whole new dimension of challenge. In this chapter we will focus on how the Four Characters behave in romantic relationships. Considering how consistent and predictable each of the characters is, it should then be relatively easy to extrapolate how each of the Four Characters would behave in emotional relationships with others, including friends and family.

Obviously it would be impossible for this chapter to be all-inclusive of the many varied ways in which the Four Characters might engage with one another. However, I hope it will shed a little light on what each of the characters find attractive in a partner, how they predictably choose to present themselves, what they need and value, and what they are looking for in the long run.

Gaining these insights into what appeals to the Four Characters in relationship should help you identify some of your own strengths and patterns, as well as help you better understand how different parts of yourself contribute to those relationships that may fall short of feeding your spirit in a healthy way. Fortunately we humans are capable of growth, and knowing what your options are for how you engage your Four Characters with others should be illuminating.

We are each the home to all Four Characters, and when we are in relationship with one another, most of us will dance in and out of our different characters in predictable ways, both positive and negative. It is completely normal for us to shift our dominant character under different circumstances, and paying attention to our own tendencies is the key to both predicting our future actions and modifying our behavior, should we decide to do so. We do not have to repeat old patterns, and we do not have to engage in relationship-sabotaging behavior.

Patterns in Partnership

If you have had much experience with dating, you have probably dated each of the Four Characters at some point. Based on current statistics, you have possibly married and divorced one or two as well. In this chapter, as we examine some of the predictable dynamics and patterns of each of the Four Characters, you will hopefully recognize some of your own behavior, as well as that of those you have been involved with.

In the beginning of a new love connection, we often feel elation when someone finds us worthy of being loved. We validate each other and share a hope that we might not need to tackle the highs and lows of life alone. But what we thought might be a good match at the onset of a commitment often sours with time and life experiences. At the end of a relationship, it is important to understand what happened, and why the coupling failed, so we can be more aware of our own patterns and perhaps choose to do things differently next time.

If you have had much experience with dating, you have probably dated each of the Four Characters at some point.

Dating through technology using platforms like Match, Bumble, OurTime, and others has made it easy for us to meet a lot of people in a short amount of time. Thinking about your Four Characters and how you relate to others in intimate relationships may help you identify more quickly when you might want to invest your time versus when you might simply want to say, "Thanks for stopping by."

If you are interested in this topic of love and the brain and would like another perspective, I am a big fan of the work of Dr. Helen Fisher. If you study her material, you will find that there is a lot of overlap in how she and I view the subject of love as it relates to the anatomy of the brain, what is going on, and who might be attracted to whom. I'm a believer that when it comes to finding a true connection, our brains and values matter.

It has long been said that opposites attract, and when we look closely at couples in love relationships, it is not unusual for one person to be more left-brain dominant and the other person more right-brain dominant. On these occasions, the pairing makes up a whole brain, and more often than not their interests, contributions, and chores fall along predictable lines. These couples often become dependent on their partner instead of developing their own opposite skills. Hopefully the insights in this chapter will help right-brain/left-brain couples *avoid* the trap of becoming dependent on each other and minimizing their own growth.

In addition, although opposites may attract at the beginning of a relationship, it is often those cute little quirks we originally found to be so adorable that later deplete our patience and grate on our nerves. I hope the insights here will smooth your path and contribute positively to your overall understanding of your own relationship experiences, past, present, and future.

Please note that as I move through this chapter, I will talk about each of the characters as though they are a specific type of person. But rest assured: I do not mean to imply that any of us is only a single character when it comes to our relationships, or any other avenue in our lives. In addition, each of our Four Characters has a right to show up and express themselves in whatever way they see fit, and it is not my intention to either undermine or undervalue any of their wants or needs.

CHARACTER 1

Imagine what it might be like to be in an emotional relationship with a Character 1. By definition this character values thinking over feeling. Character 1s may be good at providing external conveniences for living, which for some might be enough, but keeping a Character 1 focused on a passionate and loving relationship may be a challenge at times and require ongoing negotiation.

Character 1s enjoy a good spreadsheet, even when it comes to charting their life events and personal timetable. They have a preconceived notion of how their life *should* look at different stages, and sticking to the details and schedule is important. Character 1s need to define the edges of their relationship so they can both confine the risk and share with their friends and family how accurately the relationship's timeline is progressing.

As a result, the Character 1 will push to define the exact status: Are we casually dating, seriously dating, are we being exclusive, are we friends with benefits? Or are we headed toward a committed relationship, engagement, and marriage, and if so, when might that be? Character 1s are eager to put the relationship in a box and define it by asking these questions: "Who do you need me to be? Do you need me to be your protector? Do you need me to rescue you? Do you need me to be your playmate, or your breadwinner? Do you need me to be sexual with you, or co-parent with you?"

Foreseeably, Character 1s feel safe inside a predictable structure, so when a Character 1 plans an initial date, there will be a set place, program, and committed block of time. Character 1s pay very close attention to the details of how they look and smell, and they will consciously put their best self forward. Character 1s see the date as having a goal: to impress and be impressed by the other. Time has value, so the date might look a bit more like an interview for a position rather than a simple chance to hang out.

Developing something of value is important to the Character 1, so they are looking for a long-term partner they can build a life with. Although a Character 1 may date a right-brain Character 3 or 4 because they are intrigued by the excitement they may feel when they are around them, Character 1s tend to feel most comfortable and safe with the predictability of other left-brain Characters 1 and 2.

In relationship, Character 1s are comfortable with other Character 1s because, like themselves, they find other Character 1s both dependable and predictable. Two alpha Character 1s in relationship, however, must be willing to negotiate their domains of control, as alphas by definition have a mind of their own, and each needs to exercise their competency. Although Character 1s value the Character 1 skill set in another, especially in the more cooperative Soft Character 1, they may find themselves in relationship with a Character 2, who would prefer to step aside and encourage the Character 1 to take the lead.

A Character 1 dates a Character 2 by coming in with a plan, which often involves rescuing that Character 2 from a world marked by high stress and extreme levels of anxiety. The Character 1 dates a Character 2 by using its Character 1 skills of organization, confidence, strength, and endurance to help make life more predictable for the Character 2. When a Character 1 comes in and uses its authority to take control, smooth things over, and make life easier for the Character 2, the Character 2 feels protected, safe, and cared for.

CHARACTER 2

Character 2s prefer the alpha Character 1 as a long-term partner because Character 1s are predictable, dependable, and helpful. Symbiotically, Character 1s feel competent in helping others, they are good at providing for and organizing others, and they thrive when in control of the relationship. In these ways left-brain Character 1s and 2s fulfill their definition of a safe match.

Character 2s may also find common ground and comfort in the company of other Character 2s, as they often share the same trepidation that the world is a fundamentally unsafe place that is filled with emotional predators. When a Character 2 exists in a steady state of adrenal drain that can leave them feeling emotionally dull, their default emotions become distrust and fear. Both of our left-brain Characters 1 and 2 run an emotional storyline that is a zero-sum game, meaning that only one player can win when causing someone else to lose. Characters 1 and 2 keep an ongoing scorecard and always know who's one up.

Seeing life as a hard bargain, two Character 2s may emotionally pair up with one another in a *you and me against the world* victim mentality. However, this does not imply that they are happy together or that they truly like one another. Although a pair of Character 2s may moan about how unfair life is to one another—as well as to anyone else who will listen—over time their unabashed hostility tends to get the better of them. Even when they are out in public, a couple of Character 2s may loudly bully, dominate, or criticize one another and then have absolutely no idea why friends, family, or even strangers tend to avoid interacting with them.

Every once in a blue moon, a left-brain Character 1 or 2 might find themselves attracted to the excitement of a right-brain Character 3. But while left-brain Characters 1 and 2 thrive when the boundaries of an emotional relationship are well defined, the last thing a right-brain Character 3 wants is to feel boxed in. In fact, the need of the left-brain characters to define the relationship may be exactly what drives the Character 3 away. Character 3s just want to let the relationship play itself out and unfold at its own pace over time.

CHARACTER 3

Pushing the societal norm, right-brain Character 3s may run wild and sow their oats relatively late into their 30s. Although they may commit for a long weekend in the south of France, monogamy and those words "Till death do us part" feel more like a death sentence than a happy life commitment. After years of noncommitted free-spirit and indiscriminate dating, the Character 3 *player* may engage in serial monogamous relationships. But getting a Character 3 down the aisle and having them choose to stay faithful may prove to be a fight against nature.

Right-brain Character 3s are thrilled by high-energy dates because they thrive on the adrenaline rush. Character 3s are innovative and creative, so they are drawn toward variety rather than predictability, and possibility over probability. A Character 1 may find a date with a Character 3 to be really exciting, enticing, and adventurous, and it may make them feel alive, but because there is so much risk in dating the adrenaline junkie, after a little while the Character 1 may feel wrung out and desperate for something that feels more safe and tame. A Character 1 can be caught up in the rush of a Character 3, but once they become fatigued, the Character 1 might choose to go back to the predictability of a Character 1 or 2.

A Relationship Gone Bad

Here is an example of a Character 3 relationship gone bad. Character 1s are committed to reeling in their emotions, while Character 3s are all about the stimulation of the experience. Eventually the Character 1 will react to the intensity of the Character 3's recklessness with fear and respond by shifting into their emotional Character 2.

A strong Character 1 who regresses into their emotional Character 2 out of fear typically feels unsettled and becomes desperate to retreat. They do this by pumping the brakes on the relationship, thus creating space so they can return to feeling the safety of their Character 1. When a strong Character 1 shifts into their fear-based Character 2, it is not pretty, as they become controlling, hostile, and hyper-judgmental.

The once carefree and playful Character 3 perceives the withdrawal of their beloved Character 1 as a threat to the relationship, and they either remain a strong Character 3 and walk away or they, too, shift into their fear-based Character 2. The Character 3, who is now a defensive Character 2, fights for the fantasized relationship with a new storyline that reassures the Character 1 that the relationship is worth saving. If the Character 1 buys that argument and deems the relationship as necessary for their own well-being, that person's Character 2 is placing the relationship above their primary identity as a strong Character 1, and emotional pain will prevail.

The same is true for the Character 3. When a strong Character 3 values the relationship as necessary for their own emotional well-being, their Character 2 is placing the relationship above their own primary identity as a Character 3, and emotional pain will prevail.

At this point the relationship that began with a Character 1 and a happy-go-lucky Character 3 having fun has now shifted into an adversarial and codependent Character-2-versus-Character-2 standoff. What started as an intriguing and exciting relationship between two strong and healthy characters has shifted into a relationship that is now steeped in pain and fear, and with that come jealousy, envy, and discontent.

For the strong Character 1, who is now operating in the relationship as an emotionally dependent Character 2, there are two options. The Character 1 can reclaim its power, retreat, close the door, and save its own identity and sanity or it can remain in the relationship as a fear-based Character 2 and, feeling controlled by its own neediness, be miserable.

Because the Character 1 functions in the mindset that walking away is quitting, and winners never quit and quitters never win, a Character 1 may close the door on the relationship temporarily but not actually let go of the handle and permanently walk away. The emotional pain and suffering of the relationship over time becomes a predictable cycle when the Character 1's emotional Character 2 feels an urgent need to right the ship and reconnect. At this point when the Character 1 concedes and its Character 2 reopens the door to the relationship, it does so at the expense of its own integrity.

To Thyself Be True

No matter which character the Character 1 is in relationship with, once it moves into its emotional Character 2, it can either end the relationship and move back to its true self as a strong Character 1 or continue the relationship as an unhappy Character 2. Here are some of the excuses a Character 1 may use to stay in an unhealthy and compromising relationship:

- I don't want to hurt them.
- I am all they have.
- We are really good together.
- Everyone thinks we are perfect together.
- It was perfect at first.
- They aren't that bad.
- I just need to do more.
- It will change as soon as . . .
- There is nothing better out there.
- The devil you know is better than the one you don't.

The lesson here is that the minute any of our characters go against their true self and shift into their reactive and defensive Character 2 to hold on to a relationship, they lay the first brick in the wall of separation. This wall rises higher when they assign another person, place, or thing their power, and their level of resentment will be equal to the amount of power they have handed over to the other.

Once a Character 1, 3, or 4 has been dragged into the emotional pain of their Character 2, they do not stand a chance of happiness until they shift back into their primary character. Two Character 2s in dispute will never agree or find long-term peace. It is impossible. Someone must be willing to lay down their pain and shift out of their Character 2 before they can offer an olive branch for open communication, apology, negotiation, or peace.

Once one of the Character 2s who are in conflict has stepped back into their Character 1, 3, or 4, the other Character 2 will either hold on to their hostility and relentlessly chew on that bone or they, too, will let it go. If a relationship ends while someone is still grasping on to the pain of their Character 2, they may hold that grudge for decades and instantly become a Character 2 every time they are reminded about that person. For true healing to occur at the end of a relationship, both people need to step back into their primary character and allow the relationship to end on a note of kindness, forgiveness, and gratitude.

Our potential for growth as whole-brain people resides in our ability to shift out of the fear and pain of our emotional Character 2 and back into our primary Character 1, 3, or 4. But before we can make that shift, we must be able to recognize when we have been hijacked by our Character 2 in the first place. Learning how to rescue ourselves when our Character 2 comes online to fight, flee, or freeze is probably the most important skill any of us will ever learn.

If you happen to be a strong Character 2 as your primary character, you may pursue a relationship with a Character 1 because they feel predictable and safe. A primary Character 2 may choose to date a Character 3 for excitement and lighthearted fun, but it won't take long for the Character 3 to move on. A primary Character 2 might find the present-minded nature of a Character 4 an attractive option because they bring with them a sense that all is right in the world. In response, an openhearted Character 4 may reach back to a Character 2 with compassion and empathy.

The travesty for the default Character 2, however, is the realization that there is no relationship—or any other external factor, including drugs, alcohol, or other addictions—that can maintain their internal experience of peace. Whenever our Character 2 assigns its ability to experience happiness to people or other external factors, we dive deeply into a codependent relationship with those sources. No matter who I am, I cannot make you happy, sad, or even mad. We each generate our own emotions, and we are each responsible for what circuitry we run in our brain.

CHARACTER 4

A true Character 4, on the other hand, is an emotionally stable force that we all have deep inside. This character brings love to the world, and when Character 4s are attracted to someone and interested in pursuing an emotional relationship, as long as they can maintain their integrity, they act on the attraction. Character 4s observe life and relationships from the big-picture perspective, and they care about the energy that the relationship brings to them and the other. The Character 4 will ask the question, "Is this a relationship that is life giving or does it energetically drain me?"

Character 4s see beauty in everything, and even as they commit their hearts they remain invincible to the foibles of others. Character 4s value the order and organizational skills of the Character 1, while the Character 1 yearns to feel those fleeting moments of deep inner grace. Character 4s may be fascinated by the thinking prowess of a Character 1 and appreciate their mastery of details. However, unless the Character 1 is able to shift into the present moment and embody their Character 3 or 4 at least part of the time, a strong Character 4 will become bored and feel that the relationship is emotionally void of true connection.

A true Character 4 is an emotionally stable force that we all have deep inside.

In relationship, the Character 1 will ask the Character 4, "What do you need me to do?" and the Character 4 will respond by saying, "I need you to simply be." The Character 1, of course, will retort by saying, "I don't know how to be, I only know how to do. But I love you, so I will try my best." At this point the Character 1 stops the hustle of doing and begins the hustle of being. The Character 4 holds hope that a true connection will be felt by both, because it believes that the Character 1 can and will succeed if they are willing to give the process a chance and not simply be in a huge hurry to attain the end goal.

The Character 4 is secure in the knowledge that we are perfect, whole, and beautiful no matter who we are or how much money we have. The Character 4 will bring its open and loving heart to any relationship it engages in, but any left-brain character that brings its judgment, criticism, and unfulfilled expectations to the Character 4 should expect the Character 4 to shake their head and walk away.

No matter how much a left-brain Character 1 or 2 may shower a right-brain Character 4 with words of love or material gifts, the Character 4 will not feel loved until their partner becomes present with them. If the Character 4 finds itself stuck in an unhealthy relationship, they may choose to shift into their Character 1 or 3 in an attempt to feel more compatible, or they may shift into their despondent Character 2 and feel lonely. Sadly, it is all too common for us to compromise the peaceful nature of our Character 4 in order to be in relationship with others.

A Character 4 who is dating a Character 2 will be emotionally supportive of that Character 2 while insisting that the Character 2 take some responsibility for their emotional upheaval. The Character 4 will role-model a contented life that spirals with deep meaning and endless possibility, but a Character 2 can only participate in that level of bliss for a little while. Eventually the Character 2 will instinctively push away from the Character 4 to protect its internal story of the zero-sum game, where happiness only comes at a cost. To the Character 2, the big-picture, big-as-the-universe thinking of the Character 4 comes across as Pollyannaish, and for the Character 2, that level of peace is ultimately unobtainable no matter what they do.

THE MORAL OF THE STORY

If your primary Character 1, 3, or 4 has been hijacked by your Character 2 when it comes to the demands of a relationship, how do you find your way back into your healthier self? Sometimes recovering an unhappy relationship is the right answer, especially if you have two mature people who are willing to stand in the fire of their own pride, shift out of their pain, and then step back into their wholeness. However, when either of the participants is not willing to take responsibility for their Character 2, walking away and choosing one's own mental health and happiness over the relationship might be the best option.

Our potential for growth as whole-brain people resides completely in our ability to shift out of our Character 2, and we can train our healthy characters to recognize when we have been hijacked. If you find yourself in a relationship where you are not able to reclaim the position of your healthy character after you have moved into the pain of your Character 2, please go back to the chapter on the Brain Huddle. Peace really is just a thought away, and we do have the power to save ourselves when our Character 2 comes online. Practicing the huddle directly strengthens the circuitry in our brain that allows us to recover more quickly.

Now that we have explored how our Four Characters connect with our body, as well as how they connect in romantic relationship with others, let's shine a light on what happens when our brain breaks down in its ability to connect at all and how our Four Characters might strategize a successful recovery.

CHAPTER 11

DISCONNECTION AND RECONNECTION— OUR FOUR CHARACTERS IN ADDICTION AND RECOVERY

As I discussed in the earliest chapters, at the level of the single-celled organism, the meaning of life appears to be the ability to stimulate and be stimulated by that which is outside of oneself. The semipermeable membrane of the single cell allows some things to enter into the organism while blocking other things from entry. In addition, that membrane is stippled with specific types of receptors that will attract the cell toward some things in the external world or repel the cell and push it off onto a different trajectory, like two repelling magnets.

When the consciousness of the universe conceived single-celled microbes, it not only attained a higher level of order in the form of life, but it also manufactured a way in which it could keep itself stimulated and probably entertained. Via the semipermeable membrane, the universe manifested a way in which a portion of itself could be pinched off from the rest, establishing an original duality of consciousness through the establishment of a this and a that (a life-form and a universe). With the creation of the microbe, a conversation between the consciousness within the cell and the consciousness of the universe embarked full force.

This conversation is comparable to what is going on inside each of our brains, except we are multicellular creatures rather than single-celled microbes. As a result, our neurons exist in three levels of environment rather than just the inner world of the microbe and the outer world of the universe. Neurons have an internal world that is separate from and functionally in relationship with the immediate extracellular space that surrounds it. The extracellular matrix is the space that is located between the different neurons within our brain, and neuronal intercommunication is completely dependent on the molecules (and the electrical charges of those molecules) for their communication to be successful.

In the human, a brain's intelligence is dependent on the number of connections between the neurons that make up that brain. Functionally, intelligence is not simply a product of the size of a brain or the number of neurons in a brain. In order for intelligence to occur, the neurons must share their information with one another via connections.

A brain that has a lot of connections between its neurons is like a person with a computer who has access to all of the information on the Internet, compared to a person who is working with a computer that is not online. The person with the Internet connection has access to a massive volume of information, and the person without an Internet connection only has access to whatever is stored on their hard drive. Similarly, the more connections we have between our neurons, the more those neurons communicate with one another, adding their information to the overall knowledge base. In addition, as we increase the connections between the neurons in our brain, we develop higher levels of differentiation and refinement in the quality of our ability to think and feel.

CELL TRAUMA AND CONSCIOUS LIVING

When I was cut off from the thinking and emotional cells of my left hemisphere on the morning of the stroke, I was no longer privy to the information those cells held. Subsequently I lost language and any ability to comprehend that there were people who were separate from me. I lost all ability to communicate with anyone because I did not know that they existed at all.

In and of itself, losing my left-brain consciousness was unquestionably an amazing experience, but if I am to be alive and function as a normal and healthy human being in relationship with other people, my neurons are clearly a precious commodity. I have learned that my worldview is 100 percent dependent on the health and well-being of my brain cells and their connections with one another. I worked so hard for eight years to replenish those neural connections that I not only recognize their value but now do everything I can to protect them.

Yet not all of us take equal care to protect our brains' connections. Because we live in a world that values what we have in the form of external wealth and celebrity (Characters 1 and 2) over who we are as a precious form of life (Characters 3 and 4), many of us are not finding meaning in our lives and are choosing to escape through the use of alcohol and drugs.

In choosing the topic for this chapter, I certainly recognize that it would be much more fun for us to explore how the Four Characters select a Netflix series or pick a vacation. But at this juncture we need to investigate this subject, because there is nothing more destructive to our brains than addiction to drugs and alcohol. Addiction is a disease that disregards any socioeconomic or educational boundaries. It is a universal problem that does not discriminate between the homeless man and the millionaire in the mansion.

The neurological abuse that so many of us are actively engaging in through the use of alcohol and drugs is not only self-destructive but gets in the way of us having healthy relationships with other people. And it places an enormous strain on the health and well-being of humanity as a whole. A healthy society is made up of a collection of healthy brains, and a healthy brain is made up of healthy cells that are in communication with one another. We have the ability to choose whether we want to live our lives running on automatic or live our lives more consciously. We can either bounce around unconsciously like microbes in the wind with no direction, or we can choose to evolve our brains toward whole-brain living. Calling our Four Characters into the Brain Huddle offers us a tool that we can use to do this and to purposely live more consciously directed, balanced, and meaningful lives.

How we want to live is a personal decision, and I realize that many of us have chosen to use drugs and alcohol out of a desire to simply disconnect from reality, for whatever reason. Unfortunately our brains are naturally inclined to become addicted, and the more we disconnect from reality, the more our brain cells disconnect from one another and the more rigid we become in our thinking and feeling. When we are running our addiction circuitry, we are running on automatic in such a way that our circuitry is running us. This is exactly the opposite of living consciously and choosing who and how you want to be. If an addiction is running your brain, rest assured that you can find support, and there are effective tools that you can use to regain your power, break those cellular patterns, and live the life you want to live.

The best news of all is that neuroplasticity is real, and we do have the power to heal our brains and have a successful recovery if we are willing to consistently do the work. Millions of people all around the world are using the Alcoholics Anonymous Twelve Step Program as they attempt to attain and hold on to their sobriety. In the following pages we are going to look more closely at various tools as they relate to the brain, including the Twelve Steps, the Brain Huddle of the Four Characters, the Hero's Journey, and the story of the Buddha's journey to enlightenment. Although each of these tools and narratives uses different language, they all describe a similar shift in awareness and consciousness at the level of the brain.

Here we will explore the issues of addiction and recovery as they relate to the Four Characters in the hope that we might gain some insight into how we can effectively help ourselves and our loved ones who are in need.

When Addiction Touched My Life

Many years ago I found myself in a romantic relationship with an addict who was abusing both drugs and alcohol. In my naivety I forced my love to choose between having a healthy relationship with me and continuing to abuse. To my horror and surprise, I lost that gamble, and I found myself at an Al-Anon meeting wondering what the heck had just happened to my life. In that meeting I learned that although my primary relationship had been with a person, my partner's primary relationship was with alcohol. Although this was a devastating realization for me, this clarity gave me the courage to practice self-love, let go of the relationship, and choose my own mental health.

Up to that point in time, I had spent my academic career studying the brain with a focus on schizophrenia at both the neuroanatomical and psychiatric levels. My own painstaking experience with rebuilding my brain following that severe stroke left me in awe of how vulnerable and fragile both life and this beautiful organ really are. Having worked so hard to recover my own brain, it seemed completely unconscionable to me that anyone would choose to disrespect and abuse their brain cells on purpose.

As a curious scientist, once addiction had touched my life directly, I naturally began to explore the power of addiction at the level of the brain. Not only did I want to better understand what was going on at a cellular level in the brain circuitry of those who abuse drugs and alcohol, but equally important, I wanted to comprehend what was going on in the hearts and minds of those of us who love these people as our friends and family. Inevitably this line of questioning has led me into the centuries-old theme of *suffering*, and why is it that we human beings will stay in emotional relationships that are clearly not life giving? Also of equal importance, how might we support people who are not interested in helping themselves?

> **As a curious scientist, once addiction had touched my life directly, I naturally began to explore the power of addiction at the level of the brain.**

Prior to my stroke and during my 20s, I was addicted to menthol tobacco. For this reason I am intimately familiar with the stories we tell ourselves and others about why we are engaging in these destructive behaviors, e.g., the menthol opens my nasal passages so I can breathe more deeply. My favorite excuse for smoking was that cigarettes slowed my brain down enough so I could type as quickly as I could think. I *used* tobacco to help me write my dissertation. That may actually have been true, but it was still a poor excuse.

During the decade that I was a smoker, I felt incredible shame. After all, I was a medical professional in training who knew not only how hazardous smoking was for my health but how disrespectful it was to my cells. Regardless, my deep shame was not powerful enough for me to quit. I tried to stop several times, but the craving was stronger than my discipline, so I would start up again. I hated that I would binge on another pack and then have to start that counter all over again due to my relapse. I felt deep pain that here I was, a strong academic with a powerful mind, being controlled by something that was four inches long. The worst part was that once I indulged, I would crash even deeper into despair than when I was just craving. I was miserable about how deeply seated the addiction was in my brain, and I abhorred its power.

I did eventually stop cold turkey when my mother, in all of her infinite wisdom, offered me—a starving graduate student—$10 a day for every day I did not smoke, for the rest of my life. My Character 1 jumped right on that bribe, my Character 2 went into therapy about the addiction, and after celebrating three months I was so elated to be a nonsmoker that I let my mom off the hook. I am so grateful to this day for that bribe, yet even now, over 30 years later, that addiction is so engrained in my brain that I occasionally find myself smoking in my dreams.

I want to emphasize that I understand completely that addiction at the level of our brain is a powerful and devastating condition. I am absolutely in no way minimizing the experience or the profound nature of these disorders, and I respect that many of those who remain caught in the throes of any form of addiction desperately want relief and often live their lives in the ongoing fear that one day, they will use again.

A COMMON STORY

Addiction is viewed as a family disease, and when the Four Characters conversation is used within a family that is willing to stop their judgment and lay down their swords, people may better understand what is actually going on in the thought processes of the other. Al-Anon, which I mentioned above, is a special program for the friends and family of alcoholics, and it uses Al-Anon language. Alcoholics Anonymous is a program for users and its language is specific to it. By contrast, the Four Characters can offer a common language for both the user and their loved ones, offering clarity and understanding to what is actually going on in the thinking processes of both parties.

Let's listen in on an example of an internal conversation that may be going on inside the brain of an alcoholic who is intoxicated and the brain of a loved one, framed in the language of the Four Characters.

Let's say I am an alcoholic and when I am drinking, my brain is singularly focused on what it feels like to be intoxicated. When I am flying high as a kite, I am not aware that I have Four Characters because the voice of the alcohol has completely hijacked my brain, and I'm not capable of rational thought. I feel numb because my brain cells are pickled, and I am no longer capable of feeling any real emotions. By simply using my drug, I have created an enormous disconnect from my life, my pain, my Four Characters, and those who are trying to have a relationship with me.

Because of my alcohol consumption, as one would predict, the Character 1 details of my life fall through cracks. In my stupor I miss scheduled events, and my friends and family are upset because they feel, yet again, neglected and disrespected because I am not showing up as my best and sober self. With my normal range of emotions now limited, which is probably why I drank in the first place, I am emotionally removed and unavailable. My brain cells are high, and by overindulging in the alcohol, I am giving them the message that I don't value them and I don't care if they function at all. Because my brain cells are traumatized, they disconnect some of their connections with other neurons, and my thinking and emotional abilities become more rigid and closed-minded because I have fewer cells to work with.

Imagine, then, that my friends and family come to me and gently explain that we had a plan to get together, but because I was drinking, I was neither physically nor emotionally available. As an addict my own Character 1 now comes online, and in its harsh and critical judgment, I begin interrogating myself about how I could have possibly done that, once again letting down not only the people I love but myself. Inside the throes of my own indefensible behavior, my Character 1 is displeased because I let alcohol take control of me, so I did not execute the plan that it (my Character 1) had laid out for me. Instead I irresponsibly relinquished my willpower, neglected my personal hygiene, and denigrated the people I most care about. I used alcohol to turn off the voice of my Characters 1 and 2, and by doing so I completely abandoned everyone, including myself.

At this point in my internal dialogue, Character 2 pops in with a feeling of intense remorse, and I unforgivingly judge myself to be a total failure. Yet at the same time, my Character 2 delineates that I am different from those normal nonaddicts and they do not understand me at all. Consequently, I feel remorse and isolated because I am *terminally unique* and not like them. They don't understand that they can go and party and it is easy for them, but I can't do that, so I feel like I am alone in my pain.

Subsequently I not only let myself down but I let my friends and family down too. I feel shame because this is not my first rodeo, and I am well versed in how to berate myself. I fall into deep despair and wallow in hopelessness. I condemn myself for my own pathetic weakness against the drug. I feel embarrassed, and I might even hate myself for the high. Yet inside my pulsating head, my Character 2 feels like a pressure cooker that is ready to explode with vehement hostility and blame. Of course my addiction is all your fault!

But then I sober up and get some sleep, and if I'm healthy enough, my Character 3 comes back online and I feel fresh and happy again. All I want to do is play with you now, so we can make up and make it all feel better. My Character 3 is eager to heal with you. It wants to just forget about what happened, do a rerun with you, and this time I'll show up as my fun-loving, charming, innocent, and attractive Character 3 self that you love so much. And because your Character 1 wants desperately to forgive me and trust me again, you go along with the plan.

And in my mind, my Character 4 says today is another perfect day, and what will be will be. Today is a new beginning and today I will not drink. So you and I make up, and we make a new plan to go get pizza tonight. Simple as that, and all is well. Your Character 1 is a bit wary, but it trusts that you can relax now, so you head off to work. My Character 3 goes to exercise, then my Character 1 goes to work, and everything is right in the world, at least until I drink again.

As for you, my friend or family member, your Character 3 is all excited because we are going to play and eat and connect with one another like we used to. But then your Character 2 starts to fear that I am going to drink again, so you start calling me every hour to confirm that I am at work and doing okay. Which really means you are checking to see if I am sober. Then your Character 1 runs home at lunch to get rid of all the booze in your house so there is no temptation for me should I stop by there after the pizza.

But in a moment of craving desire, and remorse and shame for my past behavior, my Character 2 goes over to the pizza place early, and I binge-drink a pitcher of beer before you get there. Your Character 3 comes bouncing in all excited to see me, so my Character 2 lies, and I tell you that I only had one beer. Because your Character 1 wants desperately to believe me, you don't make a big deal out of it, and your Character 3 is happy.

We are all so pleased that everything is good again. At least until the server comes over to take our pizza order and asks if we would like another pitcher of beer. Then your Character 1 goes ballistic, lashes out with harsh negative judgment, and your Character 2 reprimands me for not having any self-respect or ability to control my drinking. Then in a flash, your Character 2 takes my behavior personally and you start to cry, and then you get up and walk out. Feeling that you have abandoned me, my Character 2 looks to find peace in yet another pitcher of beer. Not at all aware that this happened because I had abandoned myself first in the moment I chose to disconnect and use again.

In your fury and pain, your Character 1 then rationalizes that if you had just called me a little more often, or controlled my money, or my time, or my friends, or, or, maybe this would not have happened. Then your Character 2 comes online and you reproach yourself for my drinking because you didn't fix me or control me tightly enough. In your own mind, your Character 2 thrashes out that you knew you could not trust me, and then you criticize yourself for trusting me at all. At this point your little Character 2 feels abandoned, judges me negatively, and may even attack me with verbal threats or a bitter tongue. Your Character 2, reeling in her own pain, feels ashamed and powerless and takes on the blame. Knowing full well that your greatest fear is that I am going to die from this disease.

A COMMENT

This kind of dynamic happens regularly between an alcoholic and their friends and family. The alcoholic's Character 2 says, "I'm scared of you, my friends and family, because if you find out that I am using again, you will judge and criticize me, but mostly, you will stop loving me." Then their Character 1 decides that they need be deceptive and hide the fact that they are drinking. So they lie, cleverly, and at their worst they gaslight us into believing that our perceptions are wrong.

The alcoholic's Character 2 will sabotage their emotional connections in an effort to keep their cover, while their friends' and family's Character 1s try to address the problem by attending Al-Anon meetings and going to therapy. *"Fix me? Help me? What on earth makes you think that you have that power?"* professes the alcoholic.

Based on what we now know about the brain, as well as the Four Characters, it is safe to assume that the portion of our brain that is truly emotionally addicted and stays addicted to drugs or alcohol are the cells of our left and right emotional centers. What this means is that if a rehabilitation program is going to succeed, both our Characters 2 and 3 must be on board and committed to doing the emotional work. If they are not willing to participate, the rehabilitation will not deliver a lasting impact.

When an addict/alcoholic strategizes any rehabilitation effort only through the filter of their Character 1, they will jump through the appropriate hoops, clean out their blood, walk the walk, and talk the talk. To a Character 1, this is a successful recovery, even though they have completely neglected to shift their emotional relationship with their substance of choice. Remembering that we are feeling creatures who think, rather than thinking creatures who feel, although our Character 1 may help us shift our beliefs and our behavior, this is not enough for us to achieve a genuine rehabilitation. Unless the Character 1 hits rock bottom and becomes emotionally compromised as their Character 2, relapse is likely.

Alcoholism and addiction are moment-by-moment crises, and although the theme of Twelve Step Programs is *one day at a time*, the biological hook of addiction rests not only in the power of choice in the present moment (Character 3) but in the pain, guilt, and shame of our past (Character 2). To complicate matters, addiction has actually rewired our brain and possibly caused damage at the cellular level, disconnecting us from many of the things that make our life worth living. Consequently, although our Character 3 absolutely needs to participate in our recovery because we require its effort to help us make good choices in the present moment, at the core of our brain's ability to crave, it is our Character 2 who holds the key to a successful rehabilitation. Our Character 2 must be willing to participate in doing the work if recovery is going to last.

When we look more specifically at addiction and rehabilitation, if our Character 1 or 3 jumps through the hoops of our program, we may look clean on the outside. But if our Character 2 is not on board and actively participating in the process, we will relapse. Our Character 1 may come on board because they have a lot to lose, and our Character 3 might show up because it wants to be connected and does not want to feel isolated. But until our Character 2 surrenders its resentments, blame, and shame to our Character 4, there will be no spiritual awakening or true transformation.

Worthy of note, it is possible for an addict who is in rehabilitation to mimic the Character 4 of another person who appears to be finding peace and gaining success inside the program. However, although it is necessary for our Character 4 to come online and lift us above the illness, our Characters 2 and 3 must be intimately engaged in the process or relapse will inevitably occur. We are feeling creatures who think, and there is no way for us to finesse around that fact when it comes to changing any core behavior.

MORE ON FRIENDS AND FAMILY

When I first started exploring the impact of an addict's behavior on the health and well-being of their friends and family based on my own familial experience with schizophrenia, I have to say that the resemblance was uncanny. How can we adequately support someone who uses or is ill and yet minimize the negative impact of that compromised brain on how we live our own lives and nurture our own mental health?

People who are in relationship with one another tend to balance each other out. For example, if one person enjoys spending a lot of money, their partner often counterbalances that spending by being more fiscally conservative. The same is true when it comes to responsibility. If an addict is behaving in irresponsible ways, it is natural for their friends and family to shift into their responsible Character 1s to balance the dance. It is important to note that counterbalancing someone who is polarized to an extreme is no fun, and the effort feels like a burden for the Character 1. When a Soft Character 1 is cornered in a counterbalance effort, they may well shift into their Hard Character 1, adding stress to the relationship.

I certainly lived with this up close and personal in my own family. Having a brother diagnosed with schizophrenia motivated me and G. G. to team up as our Character 1s. Together we assisted my brother with his medical issues, provided a roof over his head, and kept his insanity in check just enough so that he could stay out of jail. Sometimes we succeeded and sometimes we did not. Our Character 1s managed his illness the best we could, ever complicated by having to battle against confidentiality issues with a system that was not sensitive to the positive intentions of our family. Since my brother was not able to manage his own illness, the responsibility fell to us, and in many ways the conversation listed above between the alcoholic and his family and friends was exactly the same for our family with schizophrenia.

In the case of alcoholism, the Character 1s of the friends and family desperately want to stay engaged with the person they used to know, so they persist in their efforts to help their loved one show up sober. If a Character 1 gives up on their loved one, they have to face the possibility that whatever they shared with that person was not real nor had any true meaning. This can be devastating to a Character 1 who thought they were having an intimate relationship with a person while that person was actually having their primary relationship with their drug of choice.

Once the addict/alcoholic repeatedly shows up for dates and events under the influence, friends and family turn up their Hard Character 1s and start creating strict rules. They set a household program, manage tons of details, create a perfect world, and demand that the addict attend rehab (or take their meds). Character 1s will protect the image of the happy family, make up a story to compensate for bad behavior, and perhaps even protect themselves by becoming a workaholic. Hard Character 1s often choose to travel or work more, simply because dealing with their project out in the world is easier than fixing the addict they have at home.

Friends and family live in extreme stress because they never know when the alcoholic will *use* again. A Soft Character 1, when placed under the stress of living with an addict, may well change into a Hard 1 because the Hard Character 1 is driven by their fear-based Character 2. Character 1s hide the pain and suffering of their

Character 2s in the form of self-abandonment. Friends and family become desperate to keep as much sanity in their lives as they possibly can, so they often give away their power just to keep the peace. Burying their heads in the sand like this, of course, puts the addict in control of the relationship, which bodes a predictable disaster. Yet as long as the addict is working the Twelve Step Program and trying to stay straight, the hope for the friends and family Character 1 is that the relationship will one day miraculously and spontaneously return to the way it was in the beginning.

The friends and family Character 1s know very well that negotiating with an addict is a total waste of time, but rather than admitting to failure and giving up hope, they desperately hold on to the dream. To protect themselves, the friends' and family's Character 2s will go to their Character 1s and say, "You need to get a better handle on this. We need more rules, more therapy, more frequent rehab. We also need to make more money so we can eliminate the addict's stress." And just like that, in a flash they completely enable their addict to stay at home, unemployed.

The Hard Character 1 goes along with these crazy schemes because for some families, these strategies seem to work, and it is also how they manage to keep themselves sane. To the Hard Character 1, life is like an Excel spreadsheet, and if they work just a bit harder and smarter, they will find a solution. But sadly, home has become a battleground where no one feels safe. Eventually the Hard Character 1 will wake up completely exhausted, realizing that they have inadvertently abandoned themselves by not listening to the intuition of their own Character 4.

Not wanting to give up on someone they love, or dismiss their dreams as not valid, Characters 1 and 2 hold on desperately to their hope. But after enough pain, Character 1s will throw in the towel when their Character 2s become so overwhelmed, anxious, or depressed that they feel completely powerless and defeated. For the Character 1, the tighter they hold on to the reins of their hopefulness, the more permission they give to the addict to up the ante and escalate the drama to the next level.

STORIES AND STRATEGIES TO HELP OURSELVES

There are a lot of programs that are specifically designed to help us get out of our pain and restore our cognitive connections, both within ourselves and with others. Different programs, of course, have unique appeal depending on what we believe in. There are community programs directed at helping us recover our cognitive stability, deep inner peace, or specifically our sobriety. Be it recovery from our life's challenges or recovery from substance abuse, finding our way beyond the wound and into a higher level of consciousness requires awareness, willingness, and a heartfelt commitment that is supported by an open mind.

If you believe in the doctrines of religion, then you will be attracted to programs that specifically resonate with your religious beliefs. Similarly, if you describe yourself as *spiritual but not religious*, programs that use spiritual language might appeal to you. The same is true if you are agnostic or atheist. You may find the language of science and the brain more suitable and therefore more effective in how you choose to live your best life.

Regardless of your beliefs and practices, different programs or ideologies teach pretty much the same message. When it comes to helping ourselves recover from anything, the fundamental message and steps underlying the wisdom of the Hero's Journey, for example, will naturally appeal to our Characters 1 and 2, because our left-brain characters enjoy a challenge, a quest, and a competition. The story of the Buddha (remembering that Buddhism is a practice and not a religion) uses language that appeals to our Characters 3 and 4, as our right brain is the realm of our enlightenment and redemption. Recovery as it is framed in the Twelve Step Program speaks directly to our Characters 1 and 2, as it commands an acceptance that we are powerless over our drug and we must at least be open to the idea that there is a Higher Power (a Character 4).

Although these different ideologies are unique in how they address our Four Characters, the overall intention and outcome are the same. Each of these narratives is designed to help us obtain a significant realization that will lead us to a fundamental change. Each of these narratives will lead us out of our left-brain Characters

1 and 2 and into the peaceful realm of our right-brain Character 4. The *peace is just a thought away* concept of the Brain Huddle can appeal to and empower each of our Four Characters simultaneously, leading them all to invest their energy and collaborate successfully.

We all have problems and we all suffer emotionally. In the case of the Buddha, he awakened to the understanding that the cause of our suffering is our emotional attachments. Thus when we lose things, people, titles, and freedoms that we want to hold on to, we experience emotional pain. In the case of the Hero's Journey, we hear and heed a calling to embark on a great adventure, eventually stepping out of our ignorance and into our wisdom. While in the case of the Brain Huddle, our Four Characters come together in agreement to bring our best and most authentic self forward. Finally, the Twelve Steps ground our efforts toward sobriety and recovery, literally step by step. Regardless of which of these paths we are naturally inclined to follow, or which story or strategy resonates with us best, a committed effort by all four of our characters should result in some form of resurrection, and freedom from that which roots us in our pain.

When it comes to suffering, we have cells that perform that function. We either succumb to it, linger in it, or try to move beyond it. Some of us will use an addiction in an attempt to escape suffering, but sadly, that only masks the real problem that eventually will need to be addressed. These tools, and many others that are not listed here, are quite specific in their storyline but when practiced faithfully should lead to a common outcome. Peace really is just a thought away, and the key to your brain's health will rest in your ability to find the story that speaks to you, followed by your commitment to that technique.

Regardless of the details, the steps we take to find peace remain consistent. First, we must recognize that there is a problem or a change that we want to make happen, which must be followed by a willingness to work for that change. Right out of the gate, we must be willing to step out of our left-brain ego and shift into the higher consciousness or unconsciousness of our right brain. Embarking on this part of the journey is often the most difficult step, as it requires us to recognize and admit that our small-self ego-brain must step to the side if we are to grow beyond what we currently are.

> **When it comes to suffering, we have cells that perform that function. We either succumb to it, linger in it, or try to move beyond it.**

Setting down our ego, regardless of the program, can feel like death, and our left brain will fight to maintain its own existence. It does not want to hand its control over to the unknown because doing so feels scary. In the case of the Hero's Journey, these are the monsters that it must battle. For the AA Twelve Step Program, steps 1 and 2 require that we admit that we are powerless over our addiction and that we need help from our Higher Power. To obtain enlightenment the Buddha had to step away from of all of his worldly possessions, including his knowledge, money, titles, and even the people he loved. In the language of the Four Characters, we must be willing to step out of our left-brain Characters 1 and 2 and step into the present-moment consciousness of our right here, right now, right-brain Characters 3 and 4.

Each of these narratives requires a leap of faith. We must be willing to step beyond what we have held as true and recognize that there is something greater than us that will both hold us and guide us safely into the unknown. This can be a very tall order, so it might be helpful to know that even when we do choose to set our ego aside, it is always right there, eager and ready to jump right back online at a moment's notice.

The Journey of Recovery

Throughout this book we have touched on the storyline of the Hero's Journey. When we come to explore it more deeply, it becomes obvious that these stages are closely parallel to the AA and NA (Narcotics Anonymous) Twelve Step Programs. These are spiritually based programs that are successfully worked by millions of alcoholics and drug addicts throughout the world. They are organized in such a way that they provide a detailed, step-by-step road map to a successful and ongoing recovery.

The similarity starts with the very first steps. In the language of the Four Characters, in order for a Twelve Step Program to be an effective tool for recovery, the alcoholic/addict must be willing to step out of their left-brain Characters 1 and 2 and step into the consciousness of their right-brain Character 4, or at least believe that it is out there and available.

Even the final steps of these different ideologies look the same. In the case of the hero, to complete his journey he must return to his old life, completely conscious and willing to share the wisdom of his redemption with others. For the alcoholic the resurrection is a *planned and spontaneous remission* of the illness that he must continue to nurture and maintain through his relationship with his Higher Power, and he must return to his life and *carry his message to others.*

No matter which program we choose to use, once we have established a healthy relationship with our Character 4 Higher Power/God/Infinite Being, we exist in a deep connection to all that is precious in this life. This relationship with our Higher Power innately emboldens and empowers us to continue working the program of our choice, and we will sustain our sobriety and/or our deep inner peace.

A healthy brain is a brain that is made up of healthy cells that are in connection with one another. Let us take a closer look at the details of the Twelve Step Program of Alcoholics Anonymous, as written in the format of the originating members of the Society, alongside the steps in the Hero's Journey and the experience of the Four Characters in quest of a healthy brain. It is my hope that as you explore these steps and stories, you will realize your own journey.

AA Step 1. We admitted we were powerless over alcohol— that our lives had become unmanageable.

Four Characters:

My Character 1 is a great taskmaster, and its area of expertise is taking care of the details of my life. At this point my Character 1 admits that I am powerless over this addiction, and that my life has deteriorated to a level that is not sustainable or manageable.

Hero's Journey:

In the Hero's Journey, I recognize that a change needs to occur and there is a quest to be had. I hear the dragon calling.

AA Step 2. Came to believe that a Power greater than ourselves could restore us to sanity.

Four Characters:

By this point my Character 1 has admitted and acknowledged that my problem is too big for me to fix by myself. When my Character 1 looks around at others who have had this same problem, it is clear that those who are actively working the Twelve Step Program have discovered a way to control their addiction and feel better. My Character 1 realizes that those who are engaged in a successful rehabilitation have established a *spiritual* relationship with something that is greater than themselves, and that within that relationship with their Character 4, there has been a psychic shift leading to a profound redemption. I don't completely understand this at this point, but I know I want it.

Hero's Journey:

I realize there is a quest waiting for me, and I decide that I am willing to embark on that journey because I am ready for a change. At this point I begin battling my monsters that don't want me to change, including the grip of my ego-self. I choose to gather up my courage and face my fears, heeding the call of the journey.

AA Step 3. Made a decision to turn our will and our lives over to the care of God as we understood Him.

Four Characters:

My left-brain Characters 1 and 2 have been living a pretty self-centered, egocentric life.

When I look at my life closely and I am totally honest with myself, my left-brain Characters 1 and 2 are aware that if I really do want to be sober, I need to hitch my wagon up to another wagon that is more stable and reliable than I am. Since the Twelve Step Program encourages me to embrace my Higher Power, who is a Character 4 of my own inclination and of my own making rather than the God of another, my left-brain Characters 1 and 2 can relax and feel safe enough to participate. I am willing to give *my God*, my Character 4 consciousness, the keys to drive my life because up to this point my left-brain characters have repeatedly driven the vehicle of my life right off the road into a ditch.

Hero's Journey:

For me to be willing to step out of the rational consciousness of my left-brain Character 1 and into the unknown consciousness of a Higher Power Character 4, I must first be aware that this is what I want and then be willing to follow through. As I defeat my monsters, I become released from their power to hold me down as my small and fearful self.

AA Step 4. Made a searching and fearless moral inventory of ourselves.

Four Characters:

In order for me to cleanse the path to my Higher Power Character 4, my Character 1 must take a long, hard look at both the road I have traveled and the assumptions I have lived by. Many of these assumptions have kept me bound to my own Character 2's negligent self-destruction, and her emotional wounds have played an enormous role in my demise. My Character 2's pain has pitted many of the potholes that I have fallen into and consistently forced me off the road.

Hero's Journey:

On my Hero's Quest, my Characters 1 and 2 look squarely at my life, and my Character 2 takes full responsibility for the resentments that I have built up and lived with over time. In addition, my Character 2 takes full responsibility for the blame that I have placed on others. As my left-brain characters become more secure and I become more sober, I begin to embrace the possibility, and the hope, that there is a different life path out there for me. A life that is free of this pain and addiction. A life that is free of these monsters.

AA Step 5. Admitted to God, to ourselves, and to another human being the exact nature of our wrongs.

Four Characters:

Although I have not yet met my God, the consciousness of my Character 4, I am open, willing, and ready to do what I need to do to create that relationship with my Higher Power. My Characters 1 and 2 are owning the mistakes I have made in my life, and I am actively washing my slate clean as I prepare for this journey into the consciousness of my right brain. I have seen the success that others have had because their God (Character 4) has shown up in their lives. I am committed to this process and ready to step out of my left-brain shame, guilt, and pain and into the right-brain consciousness of my Character 4 Higher Power.

Hero's Journey:

I, the hero, am embracing this stage of transformation that I must undergo before I can evolve to the next level of my being. By defeating the monsters of my past behavior that have held me back, and by admitting to others, myself, and my Higher Power the unattractive truths of my existence, I am consciously stepping beyond my pain, out of my left-brain ego-self, and into my right-brain Character 4, where I will find enlightenment.

AA Step 6. Were entirely ready to have God remove all these defects of character.

Four Characters:

My Characters 1 and 2 have taken complete responsibility for all the choices I have made, the things I have done, and the pain

I have instilled in others. My Characters 1 and 2 have made peace within myself through a true forgiveness of my own weaknesses. With an open heart, my left-brain Characters 1 and 2 recognize that the reason why I have done what I have done is because of my own deep inner pain, and now that I have acknowledged and forgiven myself for my shortcomings, I am no longer paralyzed by them, and I can move forward to create a relationship with my Character 4 Higher Power.

Hero's Journey:

I am now ready to embark on my quest. I have faced the flaws and limitations of my left-brain Characters 1 and 2 and accepted responsibility for my previous actions. I have offered myself forgiveness and cleansed myself of my wrongdoings, and I am now ready for a real and lasting change. As the hero, I am now ready to step beyond the consciousness of my left brain and into the consciousness of my right-brain Character 4 so I can exist in the peaceful and euphoric consciousness of my Higher Power.

AA Step 7. Humbly asked God to remove our shortcomings.

Four Characters:

My Characters 1 and 2 have looked deep within and taken complete responsibility for my weaknesses. At this point I humbly ask my compassionate Character 4 Higher Power to come into my heart, release my small-ego Character 2 from its pain, and heal me in a way that I could not heal myself. Held divinely in the mindset of my right-brain Character 4, I feel the deep inner peace and unconditional love of this Higher Power as it restores my hope for a new beginning.

Hero's Journey:

At this point, all of my preparations are completed and I step out of my left-brain pain into the unconsciousness of my right-brain Character 4. I am immediately filled with the wisdom of the universe and spontaneously released from my self-induced, left-brain crucifixions.

AA Step 8. Made a list of all persons we had harmed, and became willing to make amends to them all.

Four Characters:

Now that I am in communion with my right-brain Higher Power Character 4, I am on a different path that is rooted in different values. In order for me to create a new foundation that is solid, however, my Characters 1 and 2 must look back at the pitfalls of the road I traveled, make a list of where I deviated from my authentic truth, and assess whom I have wronged along the way.

I have to live in this world with others, so my Characters 1 and 2 must be willing to make amends for my past behavior and try to right my history of wrongs. It is time for me to ask for the grace and forgiveness of others so I can move down this new road in peace with their blessing. This is how my Character 1 ends the hustle of *doing* for my worthiness and steps beyond the hustle of *being* what I need to be to be worthy. At this stage I am actually engaged in relationship and *being* with the consciousness of my Character 4. The hustle of my left brain stops here, and I can relax in the peacefulness of my right-brain consciousness.

Hero's Journey:

I look at my life and acknowledge the challenges as well as the monsters that I both created and then faced down to get to this point. I have found peace in this relationship with my Higher Power, and it is now time for me to examine the amends I need to make with those I turned into monsters along the way. I need to forgive others and be forgiven by others so I can release the burden of my past and move on. It is time for me to make a list.

AA Step 9. Made direct amends to such people wherever possible, except when to do so would injure them or others.

Four Characters:

Although I have made peace with myself via my Characters 1 and 2, and opened the door to my Higher Power Character 4, the road ahead would be a lot easier for everyone, including those I have hurt, if they would accept my apology, forgive my past indiscretions, and wish me well on my new endeavor. Apologizing, and thus neutralizing the pain I have caused in others, helps me not only step out of the shame of my Character 2 but move beyond it. It is not enough for me to forgive myself for my past. I need to own it, forgive it, ask for forgiveness from others, and then genuinely release it.

Hero's Journey:

When my Character 2 stops fighting and is willing to show up and make amends with everyone it has scorned, I defeat both the real and imagined monsters that I have spent a lifetime battling. When my Character 2 is able to relax and I step into the loving consciousness of my Character 4, I gain the power to step beyond the shame of my past and into the love of my Higher Power. At this point, when I accept and forgive myself for the road I traveled, I release that past pain and open my heart to the beauty of what is right here, right now in the present moment. By embracing my Character 4, I embrace my own divine being, and I feel peace.

AA Step 10. Continued to take personal inventory and when we were wrong promptly admitted it.

Four Characters:

My Characters 1 and 2 have enjoyed dominating my life for a very long time, and they remain well versed in how to live an automated, unconscious life. It is critical that I pay attention to what is going on inside of my brain so that I purposely protect myself from reverting to my old left-brain Character 1 and 2 habits that led me into drinking in the first place.

Now that I am awake in the consciousness of our Character 4, I need to purposely nurture that relationship so that circuitry can become strong. Our brains are cells that communicate in circuits, and the more often we run a circuit, the stronger that circuit becomes. This means that all of my old addiction circuitry that I have been running for a long time remains wired in my brain. In order for me to weaken those circuits and escape their cravings and desires, I need first to get sober and then purposefully and consciously strengthen my new circuits by continuing to take an honest personal inventory about what is going on inside.

My Character 2 is the home of my expression of blame, my feelings of shame, and all of my other deep emotional pains. The cells making up my Character 2 never mature, which means that these cells will *always* have a natural propensity to rerun my old addictive patterns. It is imperative that I understand that the circuitry underlying my addiction will always be in my brain, ready to rerun at any moment. This is why I must purposely protect myself from the cravings and fears of my Character 2, especially when I am feeling hungry, angry, lonely, or tired (HALT—taught by AA programs).

Hero's Journey:

Once I have met the Character 4 consciousness of my Higher Power, my spirit is cleansed and I become filled with divine wonder. However, once I have found this connection, it is important that I nurture and strengthen my relationship with my Higher Power, because once I return to my old life, my left-brain Characters 1 and 2 will come right back online and be tempted to engage in those old behaviors.

AA Step 11. Sought through prayer and meditation to improve our conscious contact with God, as we understood Him, praying only for knowledge of His will for us and the power to carry that out.

Four Characters:

When I pay attention and consciously choose to nurture the relationship between my left-brain Characters 1 and 2 and my right-brain Character 4, I strengthen that circuitry in my brain. With practice I gain the ability to instantly step out of the consciousness of my left-brain characters and directly into the consciousness and peacefulness of my Character 4. This ability is how I own my power, and how I choose moment by moment who and how I want to be in the world.

Hero's Journey:

By this point I have found my way beyond the monsters of my past, and I have stepped into the overwhelming sense of freedom and bliss that exists at the distant end of my journey. During this spiritual awakening, I have found the deep inner peace and clarity that I yearned for, as I am now enveloped in the enlightenment of my Character 4. Not only do I feel better but I feel relieved of my pain. At this point I will either choose to return to the left-brain character life that I led before, and share my hard-won insights with others, or I will choose to keep this newfound wisdom to myself. If for some reason, once I have returned to my old life, I do not continue to run my newfound circuitry, I will relapse back into the person I was before I heeded the journey.

AA Step 12. Having had a spiritual awakening as the result of these Steps, we tried to carry this message to alcoholics [others], and to practice these principles in all our affairs.

Four Characters:

Just as the hero returns to his life with a new wisdom, so does the alcoholic who is in recovery. The alcoholic has traveled from the hell of his Characters 1 and 2 addiction to the salvation and freedom of his divine, right-brain Character 4. The alcoholic is now free to consciously live a different life, knowing full well that the circuitry of his addiction remains intact, leaving him vulnerable to a relapse at any moment. The alcoholic in recovery then returns to the world of other alcoholics who are in pain and shares with them his insights and newly found wisdom. By consciously choosing to repeat the last few steps of the program routinely, he need not weaken or flounder. Our alcoholic in recovery is the life-force power of the universe, and he can consciously choose at any moment which circuitry and which character he wants to embody.

Hero's Journey:

Just as the alcoholic who is in recovery returns to his life with his new insight, so does the hero. As I return home and choose to share my enlightenment with others who are in pain, I offer them hope for a different path and a brighter tomorrow.

THE POWER OF THE BRAIN HUDDLE, HERO'S JOURNEY, AND TWELVE STEP PROGRAMS

When I consider the profound impact of these lessons, I am reminded of one of my favorite Marianne Williamson sermons. She shares with us that we have the ability to hand our problems over to God, and although God is on our team, He does not merely jump on board and help us fight the battle. Rather, He lifts us completely above the battlefield. When I pass something up to God, when I hand Him a problem, I trust totally that He will do with it what He thinks is best. I don't just give my problem up to God for advice or micromanagement. Instead I consciously step out of the fears, judgments, and disappointments of my left brain and step into the faith of my Character 4.

I trust implicitly that God (the Infinite Being, the consciousness of the universe, my Character 4) has a bigger-picture view of my life than my left brain will ever understand. As a result, when I consciously choose to slip back into my most loving and peaceful self, I know God is handling the situation. When I choose to pass a problematic scenario up to God, I am not shirking my responsibility. I am merely shifting my perspective and choosing peace over fear and fret. This is the gift and power of the Brain Huddle, the Twelve Step Programs, and the Hero's Journey. When we work these practices with commitment and offer our trust to the power of the universe, everything changes in our lives and we feel better, because we have shifted which brain circuitry we are running.

As I shared with you in Chapter 4, when my father was 80, he flipped and spiraled a cute little Miata. He did not die that day, but I became his primary caregiver for the next 16 years. In response to his need, my Character 1 came online to protect my dad and tend to him. If you have ever been a caregiver, you know the high toll that stress can take on your peace of heart and mind. In my case, although I felt that I carried all of the responsibility, I actually had very little power over his behavior. Caught up in discontent because of his new physical limitations, Hal's disgruntled Character 2 showed up without any gratitude for my efforts, even though I was doing my best to show up as a Soft Character 1.

When my father was not pleased with the decisions I was making on his behalf, his Character 2 would growl at me. My Character 2 would then become resentful that he would choose to be hostile toward my Character 1 rather than express gratitude with his Character 4 for my efforts. I was volunteering to perform a duty that fell to me, so a little gratitude and support would have gone a long way to help me feel appreciated. You have probably had a similar experience.

During that time, that Marianne Williamson sermon saved my sanity, as it was my constant companion on daily walks. I needed desperately to find a healthy way to neutralize the hostility between me and my dad, so I gave the problem up to my Character 4 consciousness and stopped ruminating about our problems. Instead of sparring as two Character 2s, I invited my dad to a painting class so our Character 3s could have some fun together. Stepping into my Character 4 consciousness when managing my *dad issues* didn't just embolden me in the battle, it lifted me above the battlefield. Gaining this assistance outside my left-brain judgment and pain generated a path for us to stay connected in light of our changed circumstances and fragile emotions.

WE ARE ALL RECOVERING

The entire focus of this chapter has been about how we might influence and promote the health and well-being of our own brain (and brain cells) so we can create healthy connections with others, and ultimately become healthy, contributing members of society. At the level of the Four Characters, the process of recovery is the same for everyone, regardless of what promotes the disconnection, be it an addiction or an emotional pain.

Day by day we are all living our lives and facing our own unique challenges. Incorporating these tools into our daily routine and choosing to live a healthy life in which we practice the Brain Huddle when we don't *need* it strengthens that circuitry so that we have it available to us when we do need it. For those who are using the Twelve Step Program, repeating Steps 10, 11, and 12 on a regular basis mimics the internal reflection of our Four Characters when they are gathered in the Brain Huddle.

Whether we are recovering from an addiction, hurt feelings, a stubbed toe, or the loss of a loved one, we have the ability to take a *personal inventory* and reflect upon our individual journey during every moment of our lives. When we choose to step into the consciousness of our loving and compassionate Character 4, we not only feel loved and worthy of that love but we dissolve into the all-knowingness that we are that love. Our number one job as living beings is to love one another, and we do this best by first loving ourselves and then connecting with the other. In addition, when we allow each other the dignity of our own pain, we grow.

COMING UP NEXT

Ultimately this book has been about the evolution of the human brain toward whole-brain living and how we can use the Brain Huddle to purposely increase the number of cellular connections we have between each of our Four Characters. Once our Four Characters find their way into the Brain Huddle, we can then consciously shift between those different modules of circuitry, at will, and choose with ease who and how we want to be.

In the following chapter we will take a bird's-eye look at the profound impact technology has made on the evolution of our human brain over the last century. More specifically, we will explore the overall impact technology has made on the prevalence of our Four Characters as they show up in the different generations, helping explain the phenomenon of our *generational gaps*.

When we better understand and relate to the values and behaviors of those who are different from us, we can learn to connect with them by focusing on our similarities rather than our differences. Whether it be neurons in our brain, people in our family, or those who are on the opposite side of our socioeconomic or political aisle, connection takes energy and effort, and in the big picture, it enriches our lives. When we choose peace within us, and we choose peace between us regardless of our differences, our brains evolve.

CHAPTER 12

THE LAST 100 YEARS— OUR FOUR CHARACTERS AND THE INFLUENCE OF TECHNOLOGY

Our brains exist in an ongoing state of evolution that is shaped by both our nature and nurture. Perhaps unexpectedly, the development of technology has completely shifted the way in which we train our brains to learn, ultimately altering what we value and how we choose to live. It is my intention here to paint in broad and sweeping strokes a generalized portrait of the social and cultural trends that occurred in the United States over the last 100 years as they relate to how technology has impacted the expression of our Four Characters.

I will present this material chronologically through the filter of the generations. Regardless of where in the world you were brought up, generational differences are real, and the advancement of technology has perhaps made more of an impact on our brain's development than we would have assumed. When it comes to relationships, when we know better, we do better. Hopefully these insights into why and how we are different will help us navigate our personal and professional relationships with more compassion.

In preparation for this chapter, I interviewed dozens of people representing the different generations that I mention here. Fortunately, during the time of this writing a very good friend resided in an assisted-living establishment for the elderly, offering me a plethora of chatty companions from both the G.I. and Silent Generation populations. From there I reached out to my friends of all ages, and my academic colleagues opened up their classrooms so I could score some amazing conversations.

As a result, this conversation about the Four Characters and how they relate to the different generations, and the influence that technology has had on our brain over the last century, will begin with a look at the G.I. Generation who fought in World War II. I will chronologically follow that population with the Silent Generation, who were a small but significant group of people born during the difficult time between the Great Depression and World War II (1928 to 1939).

From there we will explore what life was like for the massive group of Baby Boomers who were born immediately postwar (1946 to 1964), primarily to the G.I. Generation. Next comes the smaller Generation X (born between 1965 and 1976), who were predominantly the offspring of the Silent Generation. The Baby Boomers then gave birth to the enormous population of Millennials (born between 1977 and 1996), followed by Generation X's offspring, called Generation Z (born between 1997 and 2010), who are teenagers and young adults at the time of this writing. Today's children are classified as the Alpha generation (born from 2010 on).

The specific dates that are associated with each of these generational time frames are not absolute, as they vary a little depending on which source you reference. In addition, for those who were born during the transitional years between one generation and the next, the generational group that a specific individual falls into depends on which external factors had the most impact on that individual's life.

The G.I. Generation: Character 1 is United for a Cause

The G.I. Generation was born between 1901 and 1927. They endured the social and economic devastation of World War I (1914 to 1918), the pandemic of the Spanish Flu (1918), the financial devastation of the crash of the stock market (1929), and its aftermath, the Great Depression (1929 to 1939).

By the time World War II began in 1939, this G.I. Generation had come of age. Many stopped whatever they were doing in their lives to join the workforce and support the war effort. Both overseas and in the factories here at home, the men and women of the G.I. Generation came together for a cause by organizing their efforts, learning new skills, and fighting for what they believed was right. As strong Character 1s, this generation joined forces, consolidated their means, coordinated their efforts, and melded into a home team that was willing to give up their lives for the love of their family and country. Their single objective was to fight for their freedom and literally save the world from the unconscionable threat of the Nazi regime. Thank goodness this generation stepped up to the challenge in the way they did, so we could live as free Americans.

The G.I. Generation learned left-brain skills using traditional teaching tools that included books for reading, writing, and arithmetic. However, according to the U.S. Department of Commerce, Bureau of the Census statistics listed in *120 Years of American Education: A Statistical Portrait*, in 1940 fewer than 30 percent of white males and females ages 25 and older had finished four years of high school.[3] The percentage for Black people and those of other races was less than 10 percent. What this means is that by far the majority of people in the U.S. in 1940 learned life's lessons not from books but through hands-on learning, which happens in the right brain. Therefore most folks' right brains were well trained to learn skills through either apprenticeship or other experiential strategies. Consequently, the G.I. Generation established a well-balanced economy and society, based on the values of both their Character 1 left hemisphere and their Character 4 right hemisphere.

---

[3] Thomas D. Snyder, ed., *120 Years of American Education: A Statistical Portrait* (Washington, DC: National Center for Education Statistics, U.S. Department of Education, 1993), 7–8.

The Silent Generation: Seen but Not Heard

A small group of people too young to actively participate in World War II were born between 1927 and 1945, right around the Great Depression. The years prior to World War II were a bleak time, as many families had lost their homes and their possessions, and even food was hard to come by. Follow this already arduous time with the war effort in which some 400,000 Americans would lose their lives, and education was not the social priority. As a result, as with the G.I. Generation, the majority of learning came from hands-on experience and solid labor.

An ongoing theme for those born during this period was that children should be seen and not heard, hence their classification as the Silent Generation. Compounding the theme of a need for silence, in the early 1950s Senator Joseph McCarthy roused fears in the public of anti-American sentiment, thus making it dangerous for Americans to openly speak about their thoughts, ideas, or beliefs in public. As McCarthyism grew more widespread, the Silent Generation remained cautious with their tongues, but eventually they would erupt as the fervent voices of the civil rights movement of the '50s and '60s.

The G.I. and Silent Generations at Home

Although postwar Americans around 1945 may have lost their innocence upon seeing the racism and genocide that Hitler and the Nazi regime were capable of, the surviving population of the G.I. Generation buckled down and built the robust U.S. economy of the '50s. These older Americans were a dedicated and loyal Character 1 workforce who often labored at the same company for decades. As a group they respected authority, obeyed the law, and lived conservatively. Collectively, they built an economy that heralded an unprecedented period of economic growth and prosperity, and the U.S. soon became the richest country in the world.

Along with a boom in professional and social opportunities, there was also a boom in the birth of babies, with some 77 million children born between 1946 and 1964. The social creed of the times centered around life, liberty, and the pursuit of happiness. Although these postwar Americans had witnessed the worst of humanity, their collective Character 4s prioritized home, relationships, and good old-fashioned family values. They were committed to their loved ones and wanted their children to achieve the American Dream, meaning that they could do or be anything.

In the most generalized and idealized perspective of the era, with only the radio and a little TV disturbing the peace and quiet of their daily lives, these folks had time, plenty of time to engage with life at a manageable pace. The tone of their living was set by the postwar G.I.'s Character 4s, which encouraged people to stop, pause, breathe deeply, and authentically connect with one another. Interpersonal and multigenerational relationships thrived as family gatherings took center stage.

The days ebbed slowly with a hoe-dee-doe lingering that encouraged quality time with others. Evening dinners were set on a schedule and not to be missed. It was common for neighbors to stop in, as there was an open-door policy for visitors. Dads took their kids fishing and built furniture for their Mrs. and dollhouses for their little girls, while moms delighted everyone with the favorite family recipes. Men clustered around anything mechanical to catch up on the news, and women gathered to gossip, mend, can goods, and chat about the children.

As a society Americans had a communal sense of thankfulness for their restored freedom, a gratitude for what life now was, and hope for a brighter future. Yet brewing underneath all that community and Character 4 peacefulness was an unrest of economic and social inequality between not only the different races but the different genders. The Character 2s of the G.I. and Silent Generations who had felt oppressed by the ongoing status quo may have seemed calm on the surface but now gained a voice. Their discontent would ultimately explode as the civil rights movement, followed by the women's movement that gained momentum in the '70s.

The G.I. and Silent Generations at Work

The career choices for the Four Characters during these years of postwar peace and civil upheaval became relatively predictable. The left-brain dominant Character 1 males valued money, pursued jobs that required higher education and leadership, and aggressively climbed the hierarchical ladder of society. Among other professions, Character 1 males became business executives, bankers, doctors, attorneys, politicians, military officers, engineers, accountants, advertisers, and marketers. This population thrived both socially and financially. They married and generally spawned an average of three children, depending on their religion.

The Character 3 males who bypassed formal education pursued blue-collar positions as plumbers, construction workers, bus drivers, mechanics, pipe fitters, warehouse and factory workers, tool and die men, farmers, and anything at all that was related to the construction and operation of transportation. These men were good with their hands and often served as apprentices under masters of their trade. The armed forces appealed to male Character 3s who valued an external source of structure, a sense of adventure, and a predictable paycheck.

Many Character 1 women in the '50s and '60s were successful wives and homemakers, as running a house filled with children required the ability to organize people, places, and schedules. Although it was traditional for postwar American women to marry and stay at home with their children, it is important to note that there was a special population of Character 1 women who sought an education and grew up to be financially and socially independent. Some of the best Character 1 out-of-the-home jobs for women during this period were as teachers, secretaries, stenographers, and nurses.

Although it remained the social norm for young women to marry and have children in the '50s and '60s, a record number of women in postwar America filed for divorce in the '70s. In 1963 Congress passed the Equal Pay Act, and in 1964 the Civil Rights Act. Follow these actions with the 1972 Title IX law, which protected a girl's right to receive an equal education, and the battle of the sexes commenced. For the first time in the history of the U.S., both men and women were college bound. Highly educated and competent Character 1 women began flooding the workplace and were successfully competing with their Character 1 male counterparts for positions that had previously been reserved for men.

By the '70s, women who did not pursue a college education also entered into the workforce in droves. Many became waitresses, factory workers, nurses' aides, sales reps, travel agents, stewardesses, farmhands, customer service people, and/or child caregivers.

The Character 4 of the postwar American male, who valued connection, community, family, and being of service, set and dominated the overall tone of postwar life. These Character 4 males were the big-picture thinkers who thought systematically and built an economy that reflected their values. They engaged with the postwar efforts of the Army Corps of Engineers to build up the economy, and many Character 4s became university professors who invested in the growth of bright young minds. These men were committed to creating the infrastructure of suburban America, and they functioned as the stronghold anchors of any system. At the end of the day, these men built a world that was infused with innovation, family values, and vision of a bright tomorrow.

THE BABY BOOMER GENERATION: THE AMERICAN DREAM

Following World War II, 77 million Baby Boomers were born to postwar Americans (1946 to 1964), who showered their children with all the opportunities they had not been afforded themselves. The Boomers, whom some also refer to as the "Me" Generation, received endless opportunities and were reassured that they could grow up to achieve the American Dream by having and being anything they could dream. In the big picture of life, generally speaking of course, society was rich in variation and nuance. The Boomers had more wealth than any previous generation had ever had, and they enthusiastically danced their way through life to the lyrical messaging of folk rock, the screaming guitars of rock 'n' roll, and eventually the syncopated bass lines of disco.

The oldest of the Boomers turned 18 in 1964, and while many embraced higher education and the countercultural rhetoric of the '60s, the military draft sent many Boomers unwittingly into combat in Vietnam. Some 200,000 fully fledged hippies lost themselves in hallucinogenic drugs, while another 200,000, who were too young to participate in the rampant drug culture, swooned and crooned a tune as teenyboppers to teen idols and pop music. Boomers embraced all forms of entertainment, fashion, and materialism. They were extreme consumers.

For Boomers, the death of President Kennedy in 1963 and the landing of a man on the moon in 1969 profoundly impacted their worldview. As did the Vietnam War, the return of the Vietnam vet, and the impeachment of President Nixon in 1974. Each of these events elevated major political unrest and distrust, on top of what was already brewing as the civil rights movement. Yet even amid this social restlessness, at the height of the economy in the '70s, Americans continued to believe the narrative that Boomers could, and should, achieve the American Dream. Character 1 Boomers bought homes, settled down, and followed their parents into the workplace.

The majority of the U.S. jobs available for the Boomers during the 1970s and '80s were in manufacturing, both on the assembly line and in the management office. The U.S. educational system, which had been run by the states and then taken over by the federal government in 1954, trained Boomers of all ages left-brain skills through the use of traditional left-brain teaching tools that stressed memorization of facts and details over creativity.

Boomers were the first group to place a higher value on the material reward of their Character 1, over their Character 4's value of relationships and family.

For Boomers, the percentage of the population who finished four years of high school grew between 1964 and 1980, from 50 to 70 percent for white people and 25 to 50 percent for Black Boomers and other races. Consequently, much of their learning was hands-on, resulting in a large percentage of Boomers being trained to be worker bees rather than critical or independent thinkers. Millions of Boomers of all ages, races, and genders squeezed themselves into jobs that needed to be filled, and during the '70s the values of the workforce morphed into a left-brain Character 1 culture that prioritized time on the job over quality time at home.

The culture of the '70s was all about materialism, with the explosion of the designer phenomenon and name brands. Left-brain Character 1 Boomers were willing to put 60 to 80 hours a week into the job to win the expensive watch or the vacation in Hawaii. To the Boomer, dedication to getting the job done well and earning its material rewards became more important than sleep deprivation. They learned to wear the dark circles under their eyes as their badges of honor, and they were the first group to place a higher value on the external material reward of their Character 1 over their Character 4's value of interpersonal relationships and family. It is perhaps not so surprising that the rate of divorce reached record levels during this time.

The older Americans as well as the Boomers were similarly trained by an educational system that used left-brain teaching tools to teach left-brain skills, but because the older Americans still valued their right-brain creativity and quality relationships, they built a society and economy that was more balanced between the value structures of their left-brain Character 1 and their right-brain Character 4. The older Americans had built a world using their left-brain organizational skills, but they ran their communities and homes based on the values of their right-brain Character 4.

The Boomers, who had been brought up in the kind and caring Character 4 families of the older Americans, made a choice, probably an unconscious one, to lead their lives and the world with a sense of entitlement to what they had rather than with a sense of gratitude for what they had received. When Boomers shifted to value their Character 1 over their Character 4, they embarked upon the creation of the Character 1–dominant society that we live in today.

As a result, our self-value is now measured by what we have rather than by who we are. In the big picture, our right-brain kindness, compassion, honesty, openness, and valuing of healthy relationships have been sacrificed, replaced by our left brains' race to acquire the big house, the boat, and the car. And of course, the old spouse won't do anymore so we need to get a new one.

GENERATION X: THE LATCHKEY KIDS

Following the Boomers, a relatively small but important group of people were born between the years of 1965 and 1976: Generation X (Gen X). Do you recall that record number of older American Character 1 women who filed for divorce in the '70s? Add to that group the high percentage of married couples who became double-income families as more women moved into the workforce, and you end up with a population of Gen X kids who grew up with no one around when they came home from school. These *latchkey* kids, as they would come to be nicknamed, would do their chores, finish their homework, and feed the younger kids. As a population they would develop a strong sense of Character 1 responsibility and independence at a very early age.

Although divorce was tough on the family, it turned out to be a great stimulator for the economy. Divorced women boosted the banking business by opening accounts of their own, and families bought two of everything since Mom and Dad now lived in separate places. Gen X children were shuffled back and forth between their parents' two different worlds. These kids were trained to become resilient, flexible, and Character 1 independent thinkers.

From an early age, Gen X kids were given electronics in greater measure than attention. The Speak & Spell handheld educational device was introduced to Gen X children in the late '70s, and the video game industry took hold of the world by the early '80s. These kids were brought up to be technologically savvy, and they were ever ready to conquer the next mechanical gadget. They became Character 1 masters of anything with a remote, and often taught their parents and grandparents how to program the VCR, sometimes before they were old enough to read.

Yet beyond their Character 1 technological know-how, these Gen X kids were using *right-brain* tools in the form of computers and video games to learn *left-brain* skills like reading and mathematics, and this would ultimately revolutionize the evolution of their brains. There is an enormous difference between how the left and right brains learn new material. For example, you can train your left brain in the multiplication tables through rote memory by asking what is 4 X 3, or you can train the right brain to do that same equation by showing a picture of 4 monkeys + 4 elephants + 4 ostriches. As a result of this right-brain type of training through the use of games and computers, the Gen X learned to think more spatially and visually than the Boomers, G.I., or Silent generations.

Furthermore, the education that the Gen X received was not just through the lessons of the technological tools and games they learned how to use, but via the sheer determination they displayed as they investigated how to get the technology to work in the first place. These kids became explorers and learned at an early age how to poke around and push buttons in trial and error, in an effort to figure out how to get video games to go. Earlier generations, including the Boomers, were often petrified to randomly push a bunch of buttons out of fear that they would either hurt the machine or lose their data. These two populations were completely different not only in how they felt about technology but in how they used it. Boomers were willing to learn about systems and programs, but mostly because they wanted to use the technology to their advantage. The Gen X population thrived on the technology, and beyond becoming masters of the machines, they programmed and created new uses for those machines.

What this meant at the level of the brain was a significant shift in how we were training our children to think. By the mid-'90s, the Gen X kids in the U.S. were using learning tools such as the Leap-Frog educational training system to learn how to read. Thus they were learning how to use both halves of their brains simultaneously in constructive ways. By the time the Internet went online in 1993, Gen Xers were chompin' at the bit and eager to dive full force into the exciting new world of technology.

As the Gen Xers matured, broadly speaking, they disapproved of the values of the Boomers and regarded the so-called ladder of success as the path that had wrecked their families. In the eyes of many Gen Xers, Boomers were superficial people who valued brand names and net worth over their relationships with people. Boomers cared about where they ranked in the pack, while the Gen Xers didn't want anything to do with the pack, and they resented any Boomer who tried to *buy them*.

Just as Boomers were going for the "WOW, look at how much I have" factor, Gen Xers went for the "WHOA, look at how different and unique I am" factor. Gen Xers were individuals, and the culture of the '80s and '90s was individualistic to match. Everything was big and bold, with extreme sports and huge hair, and grunge was the fashion. Rock bands flourished, and MTV was a global phenomenon that spoke directly to the hearts and minds of this generation in a way that nothing else ever had. Multicolored eye shadow complemented neon clothing, and few young people had a curfew because no one was home to police them. For this population, the video store became not only the place to hang out but a primary source of entertainment.

Even as the Gen X population developed a strong Character 1, their creative and exploratory Character 3s took sports, and everything else, to the extreme. Gen Xers were brought up on video games like Pong, Pac-Man, and their many successors, which were games that gave a reward of advancement for a job well done. As a consequence, this group of Gen Xers learned through gaming that if they created a collaboration between their ordered and independent-thinking Character 1 and their creative and innovative Character 3, they could achieve the big payoff. In this way Gen Xers cultivated their individual ability to advance, advance, advance.

By the time this computer-competent Gen X flock graduated college and entered the workforce, their Character 1s were so far advanced in independent thinking that they did not fit into the box of the Boomer world that required them to work a lot of overtime and follow a rigid routine. Instead Gen Xers created the home office and programmed their computers to automate what they perceived as antiquated worker-based systems. This population enjoyed its independence and tended to have their children later than their own parents had, and flextime at the office became an important workplace option for young Gen X mothers.

Although the Gen X generation had been brought up to believe that owning a home was a good idea, unlike their parents they tended to live paycheck to paycheck and spend their money on adventures. The Gen Xers were powerfully influenced by the assassination of John Lennon in 1980 and the shooting of President Reagan in 1981. In 1986 the explosion of the space shuttle *Challenger* shook their formative world, and the savings and loan scandal solidified their lack of trust in the system. This group of Gen Xers remained committed to the power of their Character 1 individualism rather than their collective generational strength. Consequently, this group remained cautious about committing to the financial demands of the American Dream.

Later, in 2008 when the U.S. banks were loosely offering mortgages and loans without adequate credit or collateral, some Gen Xers were caught in the loop of buying real estate and living beyond their means. When the Great Recession of 2008–2009 hit, many of these Gen Xers lost their homes, as well as the financial security of their 401(k)s. That meant many Gen Xers, already adults in their late 20s, 30s, or even beyond, had to move back in with their parents, and for the first time in decades, multigenerational cohabitation became a norm.

The Millennial Generation: All for One and One for All

The Boomers and Gen Xers gave birth to 83.5 million Millennials between 1977 and 1996. Never before has there been a more biologically based generational gap than that between the brains of the Millennials and the brains of their Boomer parents. Although the Gen Xers grew up with technology and the development of the Internet, environmentally they had to fit their whole-brain nature inside the left-brain Character 1 dominant world and workforce that the G.I. and Boomer generations had established.

On a broad scale, millions of Millennials were the first human babies to share their crib with an animatronic bear named Teddy Ruxpin. As a result, for an enormous number of Millennials, their first constant companion was an electronic, battery-operated bear who became their emotional soother and neuro-regulator. In other words, the societal norm by the time the Millennials were born was that their first significant relationship would be with electronics, and this would profoundly influence them for the rest of their lives.

From a neurological perspective, the Millennial generation was reared from birth using a technology that taught both left-brain and right-brain thinking and emotional skill sets through the right-brain techniques and learning tools of computers. The level of integration between the Millennial population and technology was relatively seamless. These kids were the first generation to use computers at home for learning as well as in school. Their left brains were trained in how to use and figure out the technology, while the games and three-dimensional teaching tools made learning fun and exciting in a way that traditional techniques of books and rote memorization could not compete with.

Thanks to the Gen Xers, the right brain–educated Millennials had the chance to grow up in an environment that was a bit more welcoming to a whole-brain way of thinking and being. Consequently, Millennials have been somewhat freer to thrive on the value structure of their Character 3s, even though they do have strong Character 1s as a backup. Because the Millennials lead with their Character 3 values, which are very different from the traditional Character 1 values of their Boomer parents, there is an interesting and unprecedented tension between these two groups in the workforce. Consequently, the G.I. and Boomer populations in the traditional workplace are struggling to figure out how to either motivate these right-brain Millennials or get them to jump through their left-brain hoops to get the job done.

While most of the left-brain Character 1 breadwinning Boomers were busy at work and rarely home, the Boomer parents who did stay home, be they male or female, wrapped their entire schedule around managing their young Millennials' social calendar by shuttling them here, there, and everywhere. These Boomer parents, who had had incredible amounts of unscheduled and unattended free time when they were young, now hovered like helicopters around their own children. These extreme measures of overprotection cultivated a group of very anxious right-brain Character 3 Millennial kids who had minimal, if any, unsupervised time of their own.

Consequently, it was virtually impossible for this population to fully develop their own individual sense of safety in the world. Congruent with their right-brain nature, many Millennials grew up feeling that they did not have any control over their lives. As adults many Millennials now function in right-brain Character 3 *packs* rather than as individual Character 1s because it feels safer for them to be a part of the collective.

As parents the Boomers had good intentions, and although many of them wanted their young Millennial children to develop their left brains and engage in healthy competition, they did not want anyone to feel either left out or less than. To compensate for and soften the potential blow of losing at events, parents, teams, and schools began giving Millennial children participation ribbons and rewards just for trying something new. Our right brain embraces the collective whole, unlike the zero-sum game of our left brain that looks for winners and losers. Boomer parents wanted their Millennial children to recognize that they had value, no matter where they fell on the continuum in the performance of the pack.

This treatment of everyone as winners further encouraged Millennial children to perceive themselves as all being equal, all being the same, and all being a part of a common collective of right-brainers. It also taught them that all they had to do was show up and make an appearance and they would be rewarded. This coddling meant that Millennials had no opportunity to develop appropriate and healthy left-brain Character 2 responses to their successes and failures. At the same time it pretty much killed their left-brain Character 1 competitive drive, which is necessary if you hope to compete in a traditional workforce.

In fact, the traditional workforce is not a comfortable place for these right-brain Millennials. Many older generation left-brain Character 1 business folks have never seen anything like them. And frankly, they have little idea how to motivate this population or get them to stick around to get the job done. To the older leadership, the Millennials don't seem hungry enough to suffer for the higher-paying jobs. This is partly true, because Millennials are not motivated by money or interested in putting in long, painful hours to make a buck. If they are not happy in their job, they will leave it and go find something they like better.

In the workplace the Millennials' commitment is to the experience of the present moment rather than to the job. They are creatives, so instead of telling them what to do, they want you to present them with a problem and then trust them to go figure out a solution. Millennials are creative wizards with technology, and they think in systems. For example, in a traditional work environment that is run by the older generations, there may be 1,000 human jobs run by 10 human managers. In the Millennial world, there might be 1,000 machines doing the work with 10 managing Millennials writing the computer code.

Since the Millennials' right-brain commitment is to the experience rather than to the company they work for, they may stick with a job for two to three years and then be off to the next experience along their journey. In left-brain companies run by the older generations, this absence of commitment to an organization is often perceived as a poor sense of loyalty or a lack of allegiance to the company. But for the Millennials who run companies themselves, this predictable and steady shift of the workforce is a good idea. They love it when new people step onto their team, as they bring new insights, ideas, and skill sets to the organization. Then when they leave, a spot on the team opens up for a new person with fresh ideas. Millennials see this transience of their population as a positive, and they clearly understand their own strengths.

True to their Character 3 values, Millennials display high levels of joy and contentment when working in groups. They like to make decisions together, but overall, their Character 2s are not well developed and therefore they are highly sensitive. As a result, Millennials tend to take criticism as a personal insult rather than as constructive guidance. Creating healthy relationships in a pull-yourself-up-by-your-bootstraps working environment does not come easily for these tender souls.

For the Millennials who are running their own businesses, they see the difference in their right-brain leadership style as one that is leading with love rather than leading with a left-brain, fear-based command-and-control mentality. Millennials lead with compassion and are charitable when people make mistakes.

When it comes to the job hunt, Millennials have guiding principles about what they want to accomplish, and they go looking for a job that fits those values. Unlike older generations Millennials are less motivated by the impact their left-brain Character 1 can make in the world and are more motivated by a job that matches their right-brain Character 3 interests and skills. Somewhere along the line, probably like all generations, these kids were told that in order to succeed and be high achievers, they needed to suffer. It just so happens that the Character 3 Millennials are the first generation to buck that system and not buy what the left-brain Character 1s are selling. Millennials want to do what they want to do, and they want to do it their way. They are not willing to stay in a job they hate and compromise themselves in the same way older generations did.

Millennials know about connection and how to work with others, and they understand completely that relationships are the nucleus of any business. When the leadership style is one of domination, relationships are fragile, but when people feel supported, they will go above and beyond to perform. Millennials have figured out that you can be loving and accountable at the same time, even in the workplace, by creating an environment in which everyone can thrive.

The Character 3 Millennials are the first generation to buck the system and not buy what the left-brain Character 1s are selling.

Millennial kids grew up in a world that was completely different from anything we had known before. In 2001 young Millennials lived through the 9/11 terrorist attacks and witnessed the sadness, depression, and fear that came over everyone they loved and respected. Millennials learned early that the world is a dangerous place, and this sense was reinforced when the stock market crashed in 2008 and many families lost both their homes and their financial security. This instability in their parents' lives further escalated Millennials' anxiety, and an unprecedented number of them are now dependent on antianxiety and/or antidepressant medications as their coping norm. On top of that, a record number of Millennials are now abusing prescription drugs, which is fueling the current opioid epidemic.

These kids were brought up with a strong sense of anxiety and peril. At the same time, they were reared to believe that they could do and be anything. It has been tough for them to realize that in our society, that is simply not true. Not only was this population helicoptered by their parents, with no opportunity to figure out the world for themselves, but they grew up with social media, which, in conjunction with the participation rewards, trained them to rely on external validation for their personal value. Boomer parents determined their self-value by how they compared with the people next door, while Millennials determine their self-value based on how many "friends, likes, and clicks" they receive on social media platforms.

Millennials are inherently attached to their technology as though it is a part of their body, and when they are separated from it, they experience intense withdrawal and their anxiety soars. Millennials get their news from social media, the CNN app, Twitter, NPR, and any other apps that have appeal. They use their technology to send brief messages through text or Twitter, and they send snippets of videos through TikTok or various other forms of messaging like Instagram. If you are an older American who relies on having phone conversations with your Millennial grandkids to keep yourself abreast of what is going on in their lives, you might want to try FaceTime, Zoom, or Skype for a quicker and easier connection.

Unlike their older counterparts, Millennials have grown up so immersed in a world of technology that they are cool with the idea that their phone apps are not only tracking them but taking and selling their data. These kids are not even bothered by knowing there are hidden cameras everywhere, and that we live in a society in which they have no real privacy. For the Millennials this is the norm that they were born into, and they are comfortable with it because for them, "It has always been this way."

Millennials are a really charming group of creatives who are artists in their hearts and true to their right-brain values. They really do care that their coffee looks like art. They are a bright group of people. Many of them watch TED talks for global awareness of what is going on in the world, and they care deeply about how they might contribute to the health and well-being of humanity. Millennials want to work for a company that works in teams, cares about the community, and offers them a day off so they can donate their time to a charitable cause. Yet their greatest reward comes when they post their good deeds on social media, as it is important that their peers know what they have been up to. The hardest thing for a Millennial is to feel as though they are alone and that they do not fit inside the social group of their choice. For a Millennial who has been brought up with the constant companionship of technology, being isolated can trigger high levels of anxiety and depression, ultimately resulting in our current rates of drug abuse and suicide. Just like neurons that need to be strongly linked into their network of other neurons, Millennials thrive when they are in a healthy connection with others.

The Gen Z Generation: Technological by Nature

Next in line, following the Millennials, are Gen Z (born 1997 to 2010), who are often the offspring of the fiercely independent-minded Gen X. These Gen Z youth are even more whole-brained and independent than their parents, for several reasons. First, these kids were reared by Gen X to have highly functional Character 1s. Second, Gen Z were taught using right-brain learning tools, resulting in strong whole-brain thinking. And third, just as Gen X had to blend their technologically savvy whole-brain thinking into the establishment of the left-brain dominant Boomers, Gen Z is blending their whole-brain thinking into the right-brain dominant world of the Millennials. As a result, Gen Z is the first whole-brain generation, both biologically and culturally.

Similar to Millennials, Gen Zs have been brought up completely enmeshed with technology from as early as their cribs, and many of them spoke the language of Google long before they spoke their native tongue. However, unlike the Millennials who thrive in packs and want to be a part of a social network, Gen Zs are more socially autonomous and prefer interacting with their technology rather than with one another.

Take that a step further to realize that Gen Z actually regards their relationship with technology as an extension of themselves, as they consciously integrate technological tools into the physiology of their daily routines. There are phone apps that will monitor their vitals, count their steps, count their breaths per minute, track their sleep, slow down their heart rate, decrease their anxiety, and distract them in just about any way that you can imagine. Phone apps will tell them what to eat, when they have met their daily limit of time spent on social media, and when it is time to go to bed—and then they have apps that will play delta wave music to help them increase the quality of their sleep.

As all of this happens and our youth become more automated and neuro-regulated by their technology, the gap between us grows. These kids, and the Alpha who will follow, are neurologically unique when compared with the traditional thinking, values, and actions of the G.I. and Boomer generations. Within a century the dominance and values of our brains have shifted, and although we have known for decades that human contact with other humans helps us build healthier neural networks, technology is causing a critical disconnect between us.

Although we may use technology to increase our frequency of communication with others, it does not bring with it that spark of human connection that stimulates our brain in a positive way. We humans are wired to be social beings, and our relationship with technology is compromising our health. Research into rates of loneliness in the different generations indicates a direct correlation between high levels of technology and higher self-reported levels of loneliness. The G.I. Generation and Boomers, who did not grow up with a constant relationship with phones, pads, and computers, self-report lower levels of loneliness than these younger generations whose lives are immersed in technology. In addition, problems with unhealthy boundaries surrounding the constant use of technology have become the number one complaint of couples and families who seek therapy. Add to that the unknown impact of electromagnetic radiation on our biological systems, and technology is beginning to look like a runaway train with no conductor onboard.

In 2001 when our whole-brain Gen Z kids were very young, or not yet even born, the U.S. experienced the societal trauma and PTSD aftermath of 9/11. Compound that with the financial crisis of 2008, when the Disney vacation dwindled into the staycation, and these kids learned quickly that the world is a dangerous place and their Character 2s have good reason to be steeped in fear. Add to that the political divisiveness that has become a part of our daily rhetoric, and it is no wonder that we are caught up in the middle of an epidemic of drug overdose and suicide, especially among these younger generations who do not feel like valuable members of our human network.

If that were not enough, place these kids in the middle of the 2020 Covid-19 pandemic and it makes sense that they tend to run a bit feral. As a result, similar to the Millennials, Gen Zs spend a lot of time running their fight/flight response and are not accumulating many possessions. Instead of settling down or buying homes, these younger populations want to remain in motion because a moving target is harder to catch.

Gen Zs are independent like their parents and value their left-brain Character 1's individuality, so much so that they have no interest in fitting themselves inside the box of the establishment. As a result, many Gen Zs are choosing to skip college altogether. This population has a tremendous amount of information available to them right at their fingertips, and they literally coexist with technology as strong Character 1s and live by the values of their Character 3s. If they need something, they order it through Amazon and have it delivered almost immediately, wherever they might be. Character 3s love the instant gratification that technology has to offer.

Gen Zs are natural-born computer coders. Many of them are earning the big bucks with very little overhead as large technology companies are now hiring their skills directly through the Internet. In fact, Gen Zs are so in demand in the world of technology that major companies including Google and Amazon have gone to the extreme of no longer requiring their employees to have earned either a B.A. or B.S. college degree.

Gen Zs are interested in high-paying jobs, and you will find them driving really nice cars and wearing the latest in monogrammed fashion. For the Gen Z, their Character 1 self-worth is reflected by what they own, but it is also important for them to be able to grab what they need and go, in the event that their Character 2 feels threatened and their Character 3 needs to bolt to another location. These are very different characteristics from the typical Millennials, who as a population tend to buy their clothes at vintage or second-hand shops and are much more inclined to make a charitable donation than to spend their money on a possession.

Although Millennials thrive on social media, Gen Zs live and breathe social media. Since their primary relationship is with their phone, iPad, or computer, it is second nature for them to stay completely on top of the cultural trends, what is cool, and what is happening now. With strong right brains, this generation is much more tolerant of different cultures, sexual orientations, races, and religions, despite the incessant hate talk they are hearing from their elders. As a group these kids feel comfortable spending time doing what they enjoy rather than doing what they *should* be doing. They are artisans, and their self-pride is based on what they create on their own. The Character 4 of the Gen Z wants to grow a beautiful garden with healthy food they can eat. They care about having clean air and water, and they want to protect our planet.

Where We Are Now

As a society we have reached a tipping point in the blend of humanity and technology. What I mean by this is that although a brain is made up of billions of cells that are in communication with one another, their magical by-product is the manifestation of our individual human consciousness. Comparably, billions of our brains communicate with one another and together we manifest the collective consciousness of humanity. Add to that the understanding that the Internet is made up of billions of computers that are linked together via the consciousness of our human brains, and we end up with a global techno-consciousness that is beyond our wildest sci-fi imaginings.

In the beginning of this liaison between humans and our computers, we humans were building and influencing the computers. However, by the advent of the Millennial and Gen Z generations, it had become commonplace for the Internet to track our Internet activity, our location and movement patterns, our food and product purchases, our financial and political interests, even our faces and our friends and familial connections. Phone apps are monitoring and collecting data about our biological systems, as well as offering us advice on how we might choose to live. The ultimate integration now rests in not only how we are giving technology the power to influence our thoughts, emotions, and physiology but how we are already practicing various forms of technological and neurological microchip implantations. This is both exciting and terrifying at the same time.

Biological systems function as a collection of negative feedback loops. For example, I feel hunger pangs in my belly, so I eat food and then the pangs go away. In this system I have a desire, I act on that desire, and when the desire is negated I feel satisfied and the system returns to rest. The beauty of a system that is based on negative feedback loops is that it can create and communicate a need, and then once that need is met the system can return to its own balance and homeostasis. While in homeostasis, the biological system can rest and refuel itself. Life thrives in health with these negative feedback loops because they use the minimal amount of energy to sound an alarm, and once the alarm is addressed, the system shuts off the power and goes back to conserving energy.

Technology, on the other hand, is a positive feedback system that never pauses or stops. The more it runs—the more you play a game or browse—the more temptations there are set into the system to increase your clicks, time, and attention. Technology runs 24/7, accelerating and wearing down our neural networks. Computers and the world of the Internet run until they break down and need to be repaired or replaced, and then the system reboots and picks up where it left off. Computers drive us to work more intensely, play games harder, and think faster. Both cognitively and emotionally, technology is exhausting our biological systems and leaving us vulnerable to its addiction.

Undeniably technology offers us conveniences, helps us become more efficient, and when used appropriately, allows us to create a healthier work-life balance. Yet the go-go mentality that technology encourages us to engage in can wreak havoc on our brain health, as well as the health of our relationships with those around us. Our brains are basically the hard drive of our life, and all day long we are compiling billions of techno-cookies from the TV, our phones and social media, our techno-driven workout schedule, and of course our computers at work.

At least on a daily basis, if not several times a day, it would behoove us to empty the trash file and reboot our brains so we can function as our optimal selves. To restore a biological system that is driven by negative feedback loops, we must hit the pause button regularly and give our brain a chance to catch up, recalibrate, and regenerate itself back to the hard reset. This is one of the reasons why sleep is so important. It's also one of the advantages of consciously choosing to engage in a Brain Huddle with our Four Characters on and off throughout the day. We have the power to choose who and how we want to be, and we have the power to help ourselves, whether we are in need or just in the mood for a refreshing moment of gratitude.

Regardless of our generational variances, in that TED talk I stated, "We are energy beings connected to one another through the consciousness of our right hemispheres as one human family. And right here, right now, we are brothers and sisters on this planet, here to make the world a better place. And in this moment we are perfect, we are whole, and we are beautiful."

CHAPTER 13

PERFECT, WHOLE, AND BEAUTIFUL

Let me begin by saying how grateful my Four Characters are that your Four Characters came along on this journey.

Although my TED talk flew around the world in a whirlwind, and continues to do so, it is important to me that the message that we are perfect, whole, and beautiful not just touch you in an 18-minute flyby. Instead, I want that message to land squarely in the fertile ground of your open heart. After writing *My Stroke of Insight*, I had no intention of ever writing another book unless I felt that I had something important to say. Then I realized that most people were not aware that we have two amygdalae, hippocampi, and anterior cingulate gyri making up two functionally separate emotional systems, one in each hemisphere, and I understood why it is so challenging for people to control their emotional reactivity. When we believe that we don't have any choices, we run on automatic. When we understand the anatomy underlying our choices, not only are we empowered to not just react but we have the ability to make informed decisions. Just as Dr. Maya Angelou affirmed, when we know better, we do better.

I love a book that makes me think. But more importantly, I love a book that helps me become more conscious and evolve into my best self. One of the beautiful things about this Four Characters framework is that when you open yourself to it, it has the power to influence every moment of your life in a profoundly positive way. It is about learning to love each of the different characters inside yourself and the Four Characters in others. I believe that if you are willing to deeply explore and apply these insights to yourself and your life, you will grow exponentially.

My guess is that by now, with the exposure you have had to the Four Characters individually as well as in the wild, you are spotting these characters in yourself and in those around you. Just knowing that there are eight characters engaged in every single interaction between two people has, I hope, clarified how you might choose to interact with others more effectively. We are each one magnificent brain with Four Characters, and we have the power to choose, moment by moment, which of those Four Characters we want to embody.

> I love a book that makes me think. But more
> importantly, I love a book that helps me become
> more conscious and evolve into my best self.

When we train our brain so we can easily shift between our Four Characters, we are constructing new neural connections between those different modules of brain cells. Using those connections to bring our Four Characters into a Brain Huddle at any moment empowers us to bring our best self forward and live our life on purpose. The evolution of humanity is an ongoing process, and we have the power to consciously direct our own development as part of that evolution. We have two beautiful cerebral hemispheres, each of which processes information in its own unique way, and I believe that bringing them together into whole-brain living is our road map to both our own deep inner peace and peace in the world.

The most predictable constant in life is change, and we are wired in our right brain to meet change by being open, expansive, flexible, adaptable, and resilient. Learning to enjoy what we have while we have it, releasing it with gratitude that we had it at all, and then choosing to celebrate what comes next is one way we might choose to live. The only thing standing in the way of our expressing our joy and resilience is the wiring in our left brain that says, "No, I don't want that because I do not feel safe." Thank goodness we have this automatic knee-jerk response to push danger away, but our Character 2 is designed to be a warning, not a way of life.

When we realize that every ability we have is dependent on cells to perform that function, we become sensitive to the idea that our brain is a highly sophisticated collection of cells, and that our emotions, experiential feelings, thoughts, and behaviors are simply cells running in circuits. We are wired to be miserable just as we are wired to feel joyful, and we have the power to choose which of these circuits we want to focus our energy on and run and for how long, as well as how we will feel about it. We can choose to nip an emotion in the bud and feel the circuit run in our body and let it dissipate after 90 seconds, or we can act it out for 90 seconds, or we can rerun that circuit into a loop of emotion that goes on and on and on for 90 minutes or 90 years.

We have the power to choose which circuitry we want to run in good times and in tough times as well. A few years ago one of my dearest friends was dying. She was young, so it was particularly heart gripping for the 18 of us who supported her during her transition. None of us really knew what we were doing, but we instinctively came together and formed a human tapestry of love to support her. It was our intention to help this beautiful young soul "get out of her body" as lovingly as possible.

The night before Kat passed, there were four of us snuggled around her on the bed. At 2 A.M. her breath became labored as her chest became congested and began to vibrate with the death rattle. In that moment I realized that for the rest of my life I would remember this as either one of the most traumatic (Character 2) or one of the most beautiful (Character 4) moments of my life. I opted for most beautiful, stepped into my Character 4, and whispered into the room, "You're okay. We're okay. You only get to die once. Enjoy the ride." In an instant, the depth of her breath shifted and the tension in the room dissipated. We all accepted the reality of what was. We shifted away from our Character 2 fear and owned the knowingness of our Character 4s. With Kat in the center, we each faced the inevitable reality of her death and embraced our ability to truly love another person to the other side. She ended up having a peaceful death, and it was an amazing grace for the rest of us because we chose for it to be exactly that.

How do we empower our Character 4 when we are caught in the pain of our Character 2? Sometimes that shift can be really hard to make, but even in the worst of times, we have the power to choose which character we want to embody. And when others are willing to work with us instead of against us, our power is immeasurable.

I always thought my mother, G. G., would live to at least age 100 because she had some serious longevity genes in her maternal line. My great-grandmother had lived to 98, and my grandmother made it to 94. In May of 2015, at the age of 88 (and only three months after my father passed) G. G. was unexpectedly diagnosed with a fast-advancing cancer that would take her life within five months. As you might imagine, my little Abby Character 2 was absolutely shattered by the mere concept of losing my "mommy." G. G. had reared me twice and been my best friend and right arm for my whole life. She had rescued me and supported my recovery from stroke, and the depth of my Character 2 pain over losing her was unbearable.

Yet at the same time, thanks to my experience with the stroke, I knew that although my mother was religiously agnostic, in that she believed that she had come from the dirt and would completely disappear and return to the dirt, I knew intimately the consciousness and power of our Character 4. As you might imagine, each of my Four Characters had its own response to what we were facing. Although my little Character 2, Abby, was emotionally distraught, my Character 1, Helen, was pleased that we had a predictable timetable such that we would be able to plan the logistics and walk this road together. My Character 3, Pigpen, was excited to take everything off the calendar and live in the joy of the present moment. And my Character 4 felt assured that although losing G. G. in this life-form would leave a tremendous void in the quiet hours of my days, I would always be able to connect with her during peaceful moments of solitude. My Character 4 assured G. G. that although she was agnostic, I believed that she would be *pleasantly surprised* when she died. She looked at me blankly and said, "I guess I'll see."

G. G.'s Character 4 decided that she wanted her remaining time here to be a true celebration of her life rather than an expression of the devastation, dread, fear, and tears that our left brains are so good at conjuring. I let her know that overall I was good with that plan, but every now and again little Abby might need her "mommy" to comfort her. We verbally negotiated, agreed to the terms, set the intention for the next few months, and ended up having a ball. Among other things, G. G. wanted to be cremated, so 35 of our nearest and dearest friends came in with food and playlists of music from G. G.'s era, and we spent an evening loving on G. G. as she shared words of wisdom while we decorated the box in which she would be cremated. I now have precious video footage of my mother dancing around her cremation "box of love" to her favorite "lift me up" song, "Stompin' at the Savoy" by clarinetist Benny Goodman.

When G. G. exhaled for the very last time, I thought my little Character 2 would break down and weep. Yet instead, to my surprise, holding my mother's now deceased hand, my Character 4 looked up into the room, smiled, and said out loud, "Mama, I'm guessing you are pleasantly surprised." And then I felt elation and kissed her goodbye. Instead of hooking into the loss and pain, I spent the next few weeks consciously weaving the essence of my mother's energy into the fabric of my DNA. I let myself cry when I felt a wave of grief come over me, and although I missed my mother's presence in my daily life and still do, I strengthened our cosmic connection in such a potent way that now when I want to chat with her, I pause and breathe deeply, knowing that she is not only right here with me but that her energy fuels every ounce of my being.

We all grieve differently, but I have learned that when I open my heart to my Character 4, and consciously open my heart and mind to the Character 4 presence of those who have gone before me, I can connect with them more easily. On those rare occasions when I feel lost in the pain of my Character 2, I find it harder to feel that connection, as if my emotional pain is actually blocking my ability to feel their presence. In this way my Character 2 is actually impeding the ability of my Character 4 to connect with that which is beyond this physical form. Feeling my Character 2 emotions is neither bad nor wrong, but it's a huge waste of time and energy if I forget to enjoy and savor these deep emotions, even the painful ones, along the way.

WHERE DO WE GO FROM HERE?

Mastering strategies that allow us to shift between our Four Characters by choice is how we own our personal power. By knowing our Four Characters and learning our patterned responses, we can take the steps we need to train ourselves to switch from one character into another. If you are willing to explore what is going on inside of your head on a regular basis, on and off throughout the day, your life, your relationships, and your world will change.

Here are some suggestions for how you might immediately begin applying this material to your daily life. The first step to choosing which circuits you want to run is to observe your current patterns of thought, emotion, and behavior. In the big picture, which of your Four Characters are strong and run on automatic already, and which ones would you like to strengthen? Paying attention to your current patterns is the perfect place to begin.

If you are willing to explore what is going on inside of your head on a regular basis, your life, your relationships, and your world will change.

---

1. When You First Wake Up (and Go to Sleep)

When I first wake up in the morning, I say thank you to those cells in my brain stem that did their job and woke me up. Then I keep my eyes closed and observe what it feels like to be alive in my body. I just lie there and feel what position my body is in, and then I assess what it feels like to be me. Did my brain wake me up on its own schedule after it finished a sleep cycle, so I feel rested and content, or did I wake up prematurely in the middle of a cycle and feel as though I did not "finish cooking"?

Keeping my eyes closed so I can more easily tune in to my internal systems, I check in with my Four Characters. Do I feel like bouncing out of bed to start my to-do list? Does Abby want to stop the day and go back to sleep, or might Queen Toad want to wrap her heart around the things we are grateful for? Pigpen may either still be asleep or deep in the imaginings of a stone sculpture. All parts of our brain do not wake up (or go to sleep) at the same time, so pay attention to which of your characters wake up first and set the tone for your morning.

I think becoming conscious of and mastering your morning routine is one of the most important gifts you can effortlessly give to yourself. When I first awake, I know that all of the cells throughout my body are listening to the conversation going on between my ears. If Character 2 perks up first and announces her discomfort, all the cells in my body tend to home in and begin pronouncing my aches and pains like a roll call. If Character 3 grabs the morning mic, the neural messages of my aches and pains may utter their presence, but they fall into the background as a part of the humdrum of my existence. Each of my Four Characters will listen to those proclamations, but the outcome is predictably unique to each of those modules.

Character 2, for example, may choose to focus on the pain, concentrating on the density of the discomfort, inadvertently making it more intense. Characters 3 and 4, on the other hand, visualize the pain as a ball of energy, and when they consciously picture that tight ball of pain slowly expanding, the pain loosens its grip and lessens. Our right brain is so grateful that we are alive that our Character 3 would say, "Thank you, pain, for reminding me that I am alive. Now, how might I hold my body so that I can feel better?" Character 4 would say, "Thank you, pain, for reminding me that I am alive. I am grateful that I am alive and able to feel this pain because it means that I still have life." Once each of my Four Characters has chimed in, I can take a Brain Huddle and consciously set the tone for the day.

Of course, taking this type of a Brain Huddle at night before going to sleep is a great way to let each of the characters quiet down, become calm, and shift into slumber mode. If you find that you cannot quiet your chatty Character 1 or fearful Character 2, you can consciously choose the expansive and all-encompassing consciousness of your Character 4. It's always right there, always available for you to tap into. Hook into that consciousness and let it turn on your delta waves for deep sleep.

2. Notice When Emotions Hit

Once I am up, I take notice of when I begin to bump up against something that bristles my emotions. I am a feeling creature who thinks, so I keep a close eye out for the bristle that I feel right before I tend to reactively trigger. If I choose to become fascinated when I start feeling prickly, curious about the charge, that is often all I need to deactivate the circuit from running. We can train ourselves to exhibit the one-two punch of first, being aware, and second, shifting away and consciously choosing to not engage.

I have noticed that my immediate physiological/anatomical response to a bristling of my emotions is multifold. In an instant, my brows furrow, my jaw shifts forward and my lips purse, my eyes shoot to the left, and then my head tilts to the right, even if what is bristling me is on my right. It's fascinating. Give it a try. Pay attention in that instant of bristle to how you physiologically respond. Once you know this pattern of yours, you can train yourself to notice the first wave of reactivity, and nip the rest of the circuit (yelling, belittling, defending, hitting) in the bud. When we walk in with our eyes wide open to our own pattern of reactivity, especially if we have noticed past patterns of irritation in relationship with certain people, we can more easily choose to do a different dance step even when the old familiar music starts to play. Unless, of course, we just want to run that delicious circuit of rant and rave. If you do choose to do that, choose to do it consciously, with the awareness that in that moment you are causing a disconnect that may have long-term consequences.

3. Notice Those Stereotypical Four Characters Moments

I tend to run a lot of stereotypical moments during my day. Character 1 is so organized and structured that she deliberately picks up my stuff, maintains order, and keeps my kitchen clean. My right-brain characters don't even notice the mess. Helen is all about getting something done, so it is simple to notice when she is out and about and busy, even when she is just percolating in the background.

If I'm feeling remotely derailed or caught in my heavier emotions, there's my little Character 2 either hurting or bristling. It is obvious when she steps in and takes the reins, so when I train myself to recognize her first few triggers, I can choose to comfort her, let her rip, or sidestep her altogether. If I choose to let my little Character 2 rip, I need to be totally conscious of the wounds I may be inflicting or triggering in those around me.

Yet if I am brimming with excitement and ready for a great adventure, feeling goofy, eager to make a mess, or just laughing out loud, I say hello to my little Character 3. And when I feel that expansiveness in my chest, and my consciousness shifts into a deep sense of gratitude for anything at all, I know I am in my Character 4.

Catch yourself in your stereotypical moments. Feel them, enjoy them, and celebrate each of them. This will strengthen that circuitry of your awareness and help you choose to shift into those modules when you need them.

4. Tune In Randomly throughout the Day

It is one thing for you to notice the stereotypical patterns of your Four Characters, but you can also purposely tune in randomly throughout the day and figure out which character you are exhibiting. Not only does this keep the Four Characters in the forefront of your consciousness but you start evaluating the behavior of your Four Characters from a more nuanced observational perspective.

5. Schedule a Daily Brain Huddle

Training our Four Characters to take a Brain Huddle is an art form. Each of the characters must be willing to participate, so practicing in moments when you do not have a need helps it become a more habitual pattern for when you do need to huddle. Practice makes perfect, and putting a Brain Huddle on the schedule will help you create and strengthen that underlying circuitry so you can turn the Brain Huddle into a habit.

6. Pay Attention to Your Patterns

Which of your characters comes out when? Who comes out on a cold and rainy day, or a warm and sunny morning? Who gets hyped up on caffeine or sugar, and how do you feel after drinking a glass of milk or eating a heavy, meat-based meal? Which character likes to go to the movies or go on walks with a friend, and who picks your late-night TV show? Who comes out when your mother-in-law calls? We can learn a lot about our Four Characters by simply paying attention to these patterns.

7. Keep a Character Log

Keeping a log of your observations will provide insight into the prevalence of your characters, the way they show up, and even the time of day when they each may tend to appear. Maybe they cycle, and maybe they don't. The more aware you become of your predictable patterning, the more transparent you will become to yourself and the easier it will be for you to pick and choose whether you want to run your old patterned response or create a new one.

8. Create a Strategic Plan for Meeting Another's Character 2

It used to be that it was not socially acceptable for us to put our Character 2 on full public display, as we had trained our more civil Character 1s to interact with one another when we had different opinions. As such, our Character 1s would sidestep our Character 2's emotional reactivity and, after taking a time-out or 90 seconds to calm down, engage in thoughtful conversations and negotiations.

But times and social norms have shifted, and it is no longer unusual for us to encounter someone else's Character 2 in public. Having a plan for how you might strategize these situations is a good idea, and knowing how your Character 2 innately responds to someone else's triggered Character 2 is the first step in knowing, observing, and transforming your own automatic response. Past behavior may be the best predictor of our future reactivity, but neuroplasticity is real, and we do have the power to consciously practice new behaviors, creating new habitual responses at the neuroanatomical level.

Remember that two Character 2s will never come to a peaceful resolution. If someone is committed to expressing their Character 2 as anger, hostility, bullying, or belligerence, then unless you decide that you want to tangle with them as your Character 2, there are a few things you can do. First, of course, is to remain calm. To do this you must be more committed to your own sense of peace (Character 4) than your desire to be right (Character 1). If you approach an angry Character 2 as your Character 1 who wants to either fix the problem or be *right* about something, expect the Character 2's resistance to grow.

If you approach an angry Character 2 with your Character 3 or 4, the Character 2 will either run through their 90 seconds of negative emotion and come out ready to interact peacefully or they will rerun their negative circuit. Realizing that you do not have the power to stop a Character 2 from expressing itself is fundamental, and choosing to recognize that this person is in deep emotional pain may offer you the edge you need to keep them from triggering your own Character 2 fear—which is of course a natural response. If your Character 1 or 2 steps in and tries to shame, guilt, threaten, or bully the other person's Character 2, this will of course only pour gas on the fire rather than help extinguish it. Although the other person's Character 2 may respond by becoming still or silent in the moment, their wound is deep. When dominated rather than validated, the energy of a Character 2's wound circuit will strengthen rather than dissipate, and the upset will fester rather than heal.

When someone is attached to expressing their Character 2 and running that circuit over and over again, it is often a good idea to choose to leave them alone so they can regain their own composure. Of course, we don't want to abandon the Character 2s of those we love, but it is critically important that Character 2s learn how to tend to their own needs through the self-soothing natures of their more mature Characters 1 or 4. When a Character 2 is triggered and inflamed in full tantrum, it is important that the adults in the room remain in their adult mode.

Peace really is just a thought away, but it may take work to create the neural habit. This is one of my favorite quotes from author and American Tibetan Buddhist Pema Chödrön: "If we want there to be peace in the world . . . we have to be brave enough to soften what is rigid, to find the soft spot and stay with it. We have to have that kind of courage and take that kind of responsibility. . . . That's the true practice of peace." The deep inner peace of your Character 4 is wired right there in your right-thinking brain, and quieting your other characters is how you soften your heart to find that soft spot.

Perfect, Whole, and Beautiful

A healthy brain is made up of billions of healthy neurons that are in communication with one another. Comparably, a healthy society is made up of billions of healthy people who are in communication with one another. The way we as a society have embraced meditation, yoga, and mindfulness over the last few decades shows how hungry we are to gain control over the spontaneous reactivity of our emotional circuitry. Now we have another set of tools, the Four Characters, that we can use.

Because we are feeling creatures who have the capacity to think, instead of running our emotional circuitry on automatic and responding reactively, we have the power to choose to push the pause button, wait for 90 seconds while the physiology of our emotions flushes out of our body, and then choose the life we want to live. If we want to balance our lives, we need to balance our brains, and turning our Brain Huddle into a habit is a great tool for that.

Our brain is the living bio-network that functions as the power source of our existence. Yet because our society is skewed to the values of our left brain—which values that which is outside of ourselves more than our whole selves—it has been impossible for many of us to find true purpose and meaning in our lives. Thanks to my stroke, I found that purpose. I unwittingly heeded the call of my own Hero's Journey: I dropped my left-brain ego, battled my monsters, entered into the realm of my right brain, and let the power of the consciousness of the universe empower my recovery. And now, here I stand, bare to the bone, sharing these insights I gained with you, inviting you to assess where you are on *your* Hero's Journey.

At the start of this book, I shared very specific messages with each of your Four Characters. Now let's see what your characters have to say to one another:

To your Character 1, your other characters say this:

> You did it. Thank you. You took the dare and now you have so much more information about the rest of us whom you so mightily corral, both inside your own head and in the world. You may not be aware of it, but the rest of us really are grateful that you are willing to do what comes naturally to you, to protect and provide for us.
>
> As you step in as the adult authority in our life, please trust we know that without you there would be no order either in our life or in the world. All of us need you, our Character 1, to thrive at what you do. We need your discipline to help us create structure in our homes, safety in our schools, and civility in our government. Your discipline, judgment, rules, and order make our world go around.
>
> Thank you for staying faithful to the task. When you become tired or distressed, please make sleep your priority. And when you rise and feel fresh, and are good and ready to go at it again, please take a few moments to let the rest of us in. Remember that we are one brain, and when you are willing to reflect with us, we can walk the world like it is our backyard, and we can be happy, whole, and united. Let us remind

you, always, that we value your efforts and we are your head cheerleaders. (Ha, pun intended.) Together we make a fantastic team, and holding a Brain Huddle is always a good idea and just a thought away.

To your Character 2, your other characters say this:

Look at you—you made it, and we love you for your willingness to hang in there and learn about the rest of us. We hope you feel seen, heard, and deeply valued. Because of your sacrifice, your willingness to step out of the cosmic flow, you are our first line of protection, defense, and offense. We need you. We love you. You are our growth edge, and when we listen to you, we face our deepest fears and gain our greatest insights into our most mysterious selves.

The gift of you, our little Character 2, is our most vulnerable and innocent self, and we affirm that you are precious beyond measure. Please know that we are doing our absolute best to heed your warnings and live our best life. May you always feel supported by the rest of us. Character 1 is readily available to protect you. Character 4 is always present loving you, and Character 3 is ready to play with you. In those moments when you feel isolated, you are not alone. We are always right here, standing beside you just beyond the fog.

To your Character 3, your other characters say this:

Yahoo, what fun, we're almost done, congratulations! You really are the joy of our life, and you bring a dimension of beauty that is far beyond our wildest dreams. Because of your curiosity, playful nature, and generosity of spirit, we thrive in heart-to-heart connections, both within ourselves and with others.

Like a big, beautiful, beaming bright neuron, you reach out unapologetically and enthusiastically, boldly taking your position within the consciousness of humanity. You are our spark plug of life, our impulse to move, and our intimate connection with others. Thank you for reading this book. Thank you for reminding us, through your mere existence, how beautiful we are and what a gift this life is. By sharing your insights with others, we as a collective brain bring our peace into the world, and the world is a better place simply because we are here.

Finally, to your Character 4, your other characters say this:

It is with the utmost gratitude that we have the privilege of sharing this consciousness with you. Because of your insights and connection with all that is, we know at the deepest level that we are perfect, whole, and beautiful just the way we are.

Now let me turn the tables and speak from each of my Four Characters to yours:

My Character 1, Helen, says this:

Thank you to all four of your characters, for your commitment to making your life, and the lives of those around you, a more congenial and compatible place. Peace really is just a thought away. We have the power to change our society as well as the overall condition of the world because the external world is the macrocosm of the microcosm going on inside of us. When we feel peace, we project peace and that peace grows. When we are willing to show up as our whole selves, we create the world we want to live in, and when we find ourselves in need, the Brain Huddle is an effective tool. So let's use it and be the people we want to be.

My Character 2, Abby, says:

I have a lot to say. I want this book to help us (all of our Character 2s) feel better, faster, when we are feeling lonely, violated, or reactively triggered. Our world is kind of a mess because we (our Character 2s) are so powerful and scary to others. It is our nature to instantaneously become hurt, offended, or angry based upon minimal amounts of information. It is natural for us to become loud, aggressive, demanding, and manipulative, and at times we will even go on the attack to push your buttons so that you will go away. We do this to hold on to our power and protect ourselves.

To all those other characters, please remember that our Character 2s are programmed to fight/flight/play dead in the instant that we are emotionally triggered. As a result, we inadvertently disconnect from not only our own other characters but from other people as well. Trust me: it is just as maddening for us that our natural response during those reactive moments is to wreak havoc in our relationships and push people away when we most desperately really want to connect but don't know how. We are more likely to react like this on automatic when we don't have strong connections with the other characters in our brain. Please, we need you to love us in spite of ourselves.

At the same time, because I am a wounded child, I don't like the idea of taking responsibility for my thoughts, emotions, or behaviors when I am upset. This whole idea of emotional and cognitive accountability makes me feel very uncomfortable and distrustful of the implication that I could do or be better. That is why I may rail against this book and question its validity, as we all know that it is absolutely delicious to just run on automatic and let ourselves rip with hostile negativity, especially when we can be anonymous.

When we practice the Brain Huddle, we (our Character 2s) feel less vulnerable to running on automatic, as it strengthens our ability to be more conscious. We are healthier and feel more connected when we know our other characters are there supporting and loving us. So please, if I am wailing or trying to pick a fight with you, remember that I do not have the ability to mature. I am a vulnerable child and I am in pain. Please don't purposely antagonize, shame, or guilt me. Please walk the higher road, and don't let me hook you into a fight. Instead, just stay calm as your adult self, and love me from afar until I can get the rest of my team into a Brain Huddle so they can rescue me. Do that for me, and I'll do my best to do that for you. Maybe we can choose more often to stop wounding each other and instead help each other heal. I would like that.

My Character 3, Pigpen, exclaims this:

YOU ARE SO TOTALLY AWESOME! YAHOO, ALL FOR ONE AND ONE FOR ALL!

And my Character 4, Queen Toad, shares this wisdom with you:

We are each so blessed to have this experience of life: this magical combination of matter and energy, transformed into the structure of a consciousness that is capable of living, moving, feeling, experiencing, and thinking. Our life is the gift of the human experience, and when it is time for the energy of our consciousness to shift away from our cellular form, although our life will extinguish and our brains will still, in those precious moments between the this and the that, the here and the there, the life and the death, the breath and the final exhale, we will clearly see how perfect, whole, and beautiful we truly are and have always been.

All those years ago, my TED talk was about me. Now it is about you:

You are the life-force power of the universe with manual dexterity and two cognitive minds. You have the power to choose, moment by moment, who and how you want to be in the world.

Right here, right now, you can step into the consciousness of your right hemisphere, where you are the life-force power of the universe. You are the life-force power of the 50 trillion beautiful molecular geniuses that make up your form, at One with all that is.

Or you can choose to step into the consciousness of your left hemisphere, where you become a single individual, a solid, separate from the flow, and separate from me.

These are the "We" of your Four Characters.

Which do you choose . . . and when?

I believe that the more time you spend choosing to run the deep inner peace circuitry of your right hemisphere, the more peace you will project into the world and the more peaceful our planet will be.

And I still think that's an idea worth spreading.

INDEX

ACKNOWLEDGMENTS

I am grateful for the wonderful network of people who are my tribe as well as the entire TED community. You have loved and supported me during the last few years as this material has transformed from a basic understanding of the Four Characters into a true paradigm shift in how we might think about psychology, consciousness, and how they relate to the underlying anatomical structure of our brain. My heartfelt thanks to each of you for all of your insights and openhearted support.

My deepest gratitude goes to Patti Lynn Polk. Thanks to you this is a much more insightful and diverse book than I would have written on my own. Your support, humor, and mastery of the material through the eyes of a primary Character 1 has not only expanded my own understanding of the Four Character dynamics, but you have helped stretch this conversation into distant and important corners of our society that I would not have wandered into on my own. Having you close by as both a sounding board and driving force, your knowledge, experience, and rumble with your own Four Characters (as well as with mine) has helped deepen my comprehension of this material in valuable ways. I am profoundly grateful for your support, love, time, energy, and overall commitment to helping me get this material out of my brain, onto the page, and out into the world.

Anne Barthel, what a blessing it has been to have you as my Hay House editor for this work. Amazingly, we managed to stay focused and engaged even as the threat of Covid raged around us. Thank you for bringing just the right amount of elegance and clarity to this project. Thank you for giving me enough line to wander where I needed to roam, and just enough tweaking to keep me on track. You did a great job wrangling both me and this material, and our time together on Zoom has been a true joy.

Thank you, Michele Gingras, for bringing an eye for detail to the manuscript in a way that helped me stay focused on the semantics, while Helene Tivemark, you were a fantastic counterbalance that consistently lifted me back into the big-picture perspective of this extremely diverse material.

Ellen Stiefler, you totally rock as my agent/attorney and I look forward to our next endeavor.

And finally, a special thanks to Reid Tracy, Patty Gift, and the entire team at Hay House. Thank you for your patience, support, and overall commitment to this project.

ABOUT THE AUTHOR

Jill Bolte Taylor spends most of her year aboard an 80-foot riverboat in a beautiful cove on a lake in the Southeast. Accompanied by her constant companions Bella and Finley, she spends her time writing, paddleboarding, rowing, and entertaining friends and family when they come out to play.

During the colder months, Jill continues to keynote, and wanders the world as though it is her backyard as she educates the public about the beauty and resiliency of our beautiful brain. She values sunshine, adventure, and deep connection with nature and others.

Website: drjilltaylor.com

Hay House Titles of Related Interest

We hope you enjoyed this Hay House book. If you'd like to receive our online catalog featuring additional information on Hay House books and products, or if you'd like to find out more about the Hay Foundation, please contact:

Hay House, Inc., P.O. Box 5100, Carlsbad, CA 92018-5100
(760) 431-7695 or (800) 654-5126
(760) 431-6948 (fax) or (800) 650-5115 (fax)
www.hayhouse.com® • www.hayfoundation.org

———

Published in Australia by: Hay House Australia Pty. Ltd.,
18/36 Ralph St., Alexandria NSW 2015
Phone: 612-9669-4299 • *Fax:* 612-9669-4144
www.hayhouse.com.au

Published in the United Kingdom by: Hay House UK, Ltd.,
The Sixth Floor, Watson House, 54 Baker Street, London W1U 7BU
Phone: +44 (0)20 3927 7290 • *Fax:* +44 (0)20 3927 7291
www.hayhouse.co.uk

Published in India by: Hay House Publishers India,
Muskaan Complex, Plot No. 3, B-2, Vasant Kunj, New Delhi 110 070
Phone: 91-11-4176-1620 • *Fax:* 91-11-4176-1630
www.hayhouse.co.in

———

Access New Knowledge.
Anytime. Anywhere.

Learn and evolve at your own pace
with the world's leading experts.

www.hayhouseU.com